Studies in Fuzziness and Soft Computing

Volume 330

Series editor

Janusz Kacprzyk, Polish Academy of Sciences, Warsaw, Poland
e-mail: kacprzyk@ibspan.waw.pl

About this Series

The series “Studies in Fuzziness and Soft Computing” contains publications on various topics in the area of soft computing, which include fuzzy sets, rough sets, neural networks, evolutionary computation, probabilistic and evidential reasoning, multi-valued logic, and related fields. The publications within “Studies in Fuzziness and Soft Computing” are primarily monographs and edited volumes. They cover significant recent developments in the field, both of a foundational and applicable character. An important feature of the series is its short publication time and world-wide distribution. This permits a rapid and broad dissemination of research results.

More information about this series at http://www.springer.com/series/2941

Pritpal Singh

Applications of Soft Computing in Time Series Forecasting

Simulation and Modeling Techniques

Pritpal Singh
Department of Computer Science
and Engineering
Thapar University
Patiala
India

ISSN 1434-9922 ISSN 1860-0808 (electronic)
Studies in Fuzziness and Soft Computing
ISBN 978-3-319-26292-5 ISBN 978-3-319-26293-2 (eBook)
DOI 10.1007/978-3-319-26293-2

Library of Congress Control Number: 2015955355

Springer Cham Heidelberg New York Dordrecht London

Printed on acid-free paper

Springer International Publishing AG Switzerland is part of Springer Science+Business Media
(www.springer.com)

Everything that comes to us that belongs
to us if we create the capacity
to receive it

By Rabindranath Tagore

I would like to dedicate this book
to my loving parents, brother, sister,
wife and baby, who make me whole

Foreword

Applications of Soft Computing in Time Series Forecasting: Simulation and Modeling Techniques is an excellent piece of work done by Dr. Pritpal Singh. It is an up-to-date and authoritative exposition of applications of soft computing in time series forecasting. Any good book must be easy-to-read and should include a balanced mixture of both theoretical and practical issues. This book can boast of having these features.

The book contains eight chapters. Among them, five chapters are dedicated for designing time series forecasting models using soft computing techniques. This book would be useful to anyone who wants to develop an understanding of soft computing techniques, such as fuzzy sets, artificial neural network and evolutionary computing, and use them in time series forecasting. I feel this book is quite useful for advance researchers in this domain. Easy-to-read approach, easy-to-understand explanation of the models, and presence of their architectures will definitely benefit the readers. All these features place this book in a unique position.

Dr. Singh has produced a masterwork and deserves our thanks and congratulations for his effort.

B. Borah
Department of CSE
Tezpur University, India

Preface

Applications of Soft Computing in Time Series Forecasting: Simulation and Modeling Techniques is written primarily for researchers in business, government, and research organizations. There are several excellent books in time series forecasting, written from very elementary to advance levels. The writers of these books have their intended audiences. This book is different from those in that it deals with practical applications of soft computing techniques (especially fuzzy sets, artificial neural network, and evolutionary computing) in time series forecasting. This book generally discusses basic theory of these techniques and their applications with various examples without complicated mathematics.

In this book, I analyze significant problems of time series forecasting in depth, starting with model formulation, architecture, basic steps, empirical analyzes, and performance measures in terms of statistical parameters to see how well the proposed models performs.

The present book has been accomplished at the Tezpur University (Tezpur, India) and Thapar University (Patiala, India). Valuable suggestions and comments were provided by Dr. B. Borah, Department of Computer Science and Engineering, Tezpur University.

All experiments were conducted at the Department of Computer Science and Engineering, School of Engineering, Tezpur University. The empirical results presented in the book were published by esteemed journals.

Patiala
April 2015

Pritpal Singh

Acknowledgments

Prima facea, I am grateful to the God for the good health and well-being that were necessary to complete this book.

I would like to thank my supervisor, Associate Professor B. Borah of Tezpur University for giving me an opportunity to work under him on this challenging and burning topic and providing me ample guidance and support through the course of this research. I am also thankful to him for inspiring my interests in soft computing and time series forecasting. I am grateful for his guidance, encouragement, and support throughout my doctoral research work at Tezpur University and also in my life. This book comes from his helpful discussions, supervision, and meticulous attention to details. I am also grateful to Dr. Leontina Di Cecco, Springer Verlag GmbH, for her valuable comments and suggestions.

I take this opportunity to express my appreciations to all of the Department faculty members for their help and support. I am also thankful to all my friends, especially Hasin A. Ahmed, Krishna Das, J. Binong, and Navajit Hazarika for their direct or indirect help, inspiration, and motivation. I am grateful to all the technical and non-technical members of the department for their support.

I want to express my greatest gratitude to my dear mother and father, my beloved wife (Manjit) and baby (Simerpreet), my only loving sister (Neha), and brother (Baldev) for their endless love, constant support, encouragement, and patients throughout my research without which I would not have reached this position.

Contents

Abbreviations and Symbols

ACF	Autocorrelation Function
AFER	Average Forecasting Error Rate
ANN	Artificial Neural Network
BPNN	Backpropagation Neural Network
CN	Composite Neuron
EBP	Error Backpropagation
EC	Evolutionary Computing
FCM	Fuzzy C-Mean
FFNN	Feed-forward Neural Network
FL	Fuzzy Logic
FLR	Fuzzy Logical Relationship
FLRG	Fuzzy Logical Relationship Group
FTS	Fuzzy Time Series
FWDT	Frequency-Weighing Defuzzification Technique
GA	Genetic Algorithm
GP	Genetic Programming
GR	Generalizes Regression
IBDT	Index-Based Defuzzification Technique
IMD	Indian Meteorological Department
ISMR	Indian Summer Monsoon Rainfall
LEM2	Learning From Example Module 2
LPA	Long Period Average
MBD	Mean-Based Discretization
MLFF	Multi-Layer Feed-Forward
MLR	Multiple Linear Regression
MSE	Mean Square Error
PNN	Probabilistic Neural Network
PSO	Particle Swarm Optimization
RBF	Radial Basis Function
RMSE	Root Mean Square Error
RPD	Re-partitioning Discretization

RS	Rough Set
SC	Soft Computing
SLFF	Single-Layer Feed-Forward
SOFM	Self-Organizing Feature Maps
SPI	Standardized Precipitation Index
TAIEX	Taiwan Stock Exchange Capitalization Weighted Stock Index
TAIFEX	Taiwan Futures Exchange
$\bar{A}$	Mean
U	Theil's Statistic
TS	Tracking Signal
DA	Directional Accuracy
δ_r	Evaluation Parameter
R	Correlation Coefficient
R^2	Coefficient of Determination
PP	Performance Parameter

About the Author

Pritpal Singh received Ph.D. degree in Computer Science and Engineering from Tezpur University (A Central University), Tezpur, Assam (India), in February 2015. He received his Master Degree in Computer Applications from Dibrugarh University, Assam (India), in 2008. From July 2009 to June 2013, he was a Senior Research Fellow in the Department of Computer Science and Engineering, Tezpur University (A Central University), Tezpur, Assam (India). He has been appointed as a Faculty at the School of Mathematics and Computer Applications, Thapar University, Punjab (India), in July 2013. His research interests include soft computing (fuzzy set, rough set, soft set), big data analysis and modeling (especially time series analysis and forecasting), and bio-inspired algorithm and optimization. He has published many research articles in referred journals and conference proceedings. He has published two book chapters and one book in Springer. His research articles can be found in Knowledge and Information Systems (Springer), Knowledge-Based Systems (Elsevier), International Journal of Approximate Reasoning (Elsevier), Engineering Applications of Artificial Intelligence (Elsevier), International Journal of Machine Learning and Cybernetics (Springer), Stochastic Environmental Research and Risk Assessment (Springer), among others.

Chapter 1
Introduction

The best way to predict your future is to create it.
Peter F. Drucker

1.1 Introduction

As the application of information technology is growing very rapidly, data in various formats have also proliferated over the time. One category of such data is time series data. A time series is a sequence of numerical values recorded over a period, measured typically at successive points in time, usually spaced at uniform intervals—daily, weekly, quarterly, monthly or yearly. For examples, supermarkets are storing their daily sales figures, meteorology department is recording daily maximum and minimum temperatures, stock markets are preserving the daily opening and closing prices. Similarly monthly inflation figures, annual population growth etc. are recorded by government departments. Simply speaking a time series is a sequence of historical data collected at regular intervals. However, such data are useless unless they are analyzed and utilized.

Time series analysis is an important tool for forecasting the future on the basis of past history (Mahnam and Ghomi 2012). Forecasting is an essential aid to decision making and planning for effective management of modern organizations. Sales forecasting always plays a prominent role in business operation. It provides organizations reliable guidelines to project costs and allocate budget for an upcoming period of time. Scientists also face crucial challenge of arriving at accurate prediction of events like temperature, rainfall, economy growth, etc. No one can accurately predict the future, but a lot of benefit can be derived from obtaining a picture of the future. Predicting the future, we can remain prepared for it.

A forecast is an estimate of values of uncertain future events. Forecasting methods can be classified as qualitative or quantitative. Qualitative methods generally involve the use of expert judgment of experienced persons to develop forecasts and are

P. Singh, *Applications of Soft Computing in Time Series Forecasting*,
Studies in Fuzziness and Soft Computing 330, DOI 10.1007/978-3-319-26293-2_1

subjective in nature. They do not rely on any rigorous mathematical computations. Such methods are generally used when historical data on the variable being forecast is scarce. Quantitative forecasting methods are objective in nature. They rely on mathematical computations. Such methods can be used when quantifiable past data about the variable being forecast are available and the pattern is expected to continue into the future. Time series and regression are two important methods of this category.

Modern time series forecasting methods are based on the idea that history repeats itself. The recorded past values of the variable including the present value is called a time series. The time series is analyzed to discover a pattern and the series is then extrapolated into the future based on the pattern. Finding out patterns and extrapolating future events based on the patterns constitute the main subject matter of time series analysis. Such methods are generally used when much information about the generation process of the forecasted variable is not available and when other variables also do not provide any clear explanation of the studied variable. Time series forecasting is a growing field of interest playing an important role in many practical fields such as economics, finance, marketing, planning, meteorology and telecommunication.

In the modern competitive world, government and business organizations have to make the right decision in time depending on the information at hand. As large amounts of historical data are readily available, the need of performing accurate forecasting of future behavior becomes crucial to arrive at good decisions. Therefore, demand for definition of robust and efficient forecasting techniques is increasing day by day. A successful time series forecasting depends on an appropriate model fitting. Enhancing the robustness and accuracy of time series forecasting models is an active area of research. Multi-step forecasting is still an open challenge in time series forecasting. Many techniques for time-series forecasting have been developed assuming linear relationships among the series variables. A review of the literature on time series forecasting can be found in the article written by Gooijer and Hyndman (Gooijer and Hyndman 2006).

The book is dedicated to the development of time series forecasting models with increased accuracy levels. Existing time series forecasting methods generally fall into two groups: classical methods, which are based on statistical/mathematical concepts, and modern heuristic methods, which are based on algorithms from the field of soft computing (SC). For this purpose, various SC methodologies such as fuzzy set, ANN, RS and EC are studied, and it is found that fuzzy set methodology is widely used technique in this domain. The application of the fuzzy set in time series forecasting is referred to as "FTS". Hence, the propagation of the book work is initiated by providing an introduction to the FTS concepts in time series forecasting. In many cases problems can be resolved most effectively by integrating other SC techniques into different phases of the FTS models. Remaining part of the book presents five proposed time series forecasting models. First four models are based on the FTS modeling approach and its hybridization with other SC concepts, whereas the last model is based on the ANN modeling approach.

Time series analysis and its prediction itself involve tedious activities, such as their preprocessing, their transformation to identify suitable input predictor that can enhance the prediction, adjustment of various parameters associated with models, etc. Despite the fact that book describes application and development of techniques especially for weather and financial data, most of the techniques can also be applied to other domains, such as predictions of university enrolment, tourism demand, economic growth, and so on.

1.2 Issues in Time Series Forecasting

Some major issues in time series forecasting are discussed as follows:

(a) **Models**: Is it possible to predict the time series values in advance? If it is so, then which models are best fitted for the data that are characterized by different variables?
(b) **Quantity of Data**: What amount of data (i.e., small or massive) needed for the prediction that fit the model well?
(c) **Improvement in Models**: Is there any possibility to improve the efficiency of the models? If yes, then how could it be possible?
(d) **Factors**: What are the factors that influence the time series prediction? Is there any possibility to deal with these factors together? Can integration of these factors affect the prediction capability of the models?
(e) **Analysis of Results**: Are results given by the models statistically acceptable? If it is not, then which parameters are needed to be adjusted which can influence the performance of the models?
(f) **Model Constraints**: Can linear statistical or mathematical models successfully deal with the non-linear nature of time series data? If it is not possible, then what are the models and how are they advantageous?
(g) **Data Preprocessing**: Do data need to be transformed from one form to another? In general, what type of transformation is suitable for data that can directly be employed as input in the models?
(h) **Consequences of Prediction**: What are the possible consequences of time series prediction? Are there advance predictions advantageous for the society, politics and economics?

All these issues indicate the need for intelligent forecasting technique, which can discover useful information from data. The term "SC" refers to the overall technique for designing intelligent or expert system. It has been widely used in machine learning, artificial intelligence, pattern recognition, uncertainties and reasoning. More detail discussion on SC techniques is provided in the next chapter.

1.3 Research Problems in Time Series Forecasting

The FTS modeling approach is an interminable and an arousing research domain that has continually increased challenges and problems over the last decade. In this section, we present various research problems and trends associated with the FTS modeling approach. These discussions are based on the recent research articles[1] published by the author of the book. One research Problem (Singh and Borah 2013d), which is associated with forecasting the summer monsoon rainfall in India, solved using ANN, is also presented in this section. All these research problems are explained below.

Problem 1.3.1 (*Lengths of intervals*). For fuzzification of time series data set, determination of lengths of intervals of the historical time series data set is very important. In case of most of the FTS models,[2] the lengths of the intervals were kept the same. No specific reason is mentioned for using the fixed lengths of intervals. Huarng (Huarng 2001a) shows that the lengths of intervals always affect the results of forecasting.

$$\begin{aligned} F(t=1) \quad & A_i \rightarrow A_i, \\ F(t=2) \quad & A_k \rightarrow A_j, \\ F(t=3) \quad & A_i \rightarrow A_i, \\ F(t=4) \quad & A_i \rightarrow A_j. \end{aligned} \tag{1.3.1}$$

Problem 1.3.2 (*Ignorance of repeated FLRs*). After generating the intervals, the historical time series data sets are fuzzified based on FTS theory. Each fuzzified time series values is then used to create the FLRs. Still most of the existing FTS models ignore the repeated FLRs. To explain this, consider the four FLRs at four different time functions, $F(t = 1, 2, 3, 4)$ as shown in Eq. 1.3.1. Three FLRs at functions $F(t = 1)$, $F(t = 3)$ and $F(t = 4)$ have the same fuzzy set, (A_i), in the previous state. Hence, these FLRs can be represented in the following FLRG as:

$$A_i \rightarrow A_i, A_j. \tag{1.3.2}$$

Since existing FTS models do not consider the identical FLRs during forecasting. They simply use the FLR as shown in Eq. 1.3.2 by discarding the repeated FLRs in the FLRG.

The ignorance of repeated FLRs in the FLRG is not properly justified by these models (e.g., refer to some articles.[3]) Since, each FLR represents frequency of occurrence of the corresponding event in the past. For example, in Eq. 1.3.1, $A_i \rightarrow A_i$ occurs two times and $A_i \rightarrow A_j$ appears only once. These occurrences represent the patterns of historical events as well as reflect the possibility of appearance of these

[1]References are: (Singh and Borah 2012, 2013a, b, c).

[2]References are: (Chen 1996; Huarng 2001a; Hwang et al 1998; Song and Chissom 1993a, 1994).

[3]References are: (Chen 1996; Huarng 2001b; Huarng et al 2007; Hwang et al 1998).

type of patterns in the future. If we simply discard the repeated FLRs, then there is a chance of information loss. This can have impact on robustness as well as on the effectiveness of the model. Hence, to utilize more information, the following approach can be adopted to represent the FLRs of Eq. 1.3.2 in FLRG as:

$$A_i \rightarrow A_i, A_i, A_j. \tag{1.3.3}$$

Problem 1.3.3 (*Equal importance to FLRs*). In existing FTS models, each FLR is given equal importance, which is not an effective way to solve real time problems, because each fuzzy set in the FLR represents various uncertainty involved in the domain. According to Yu (2005), there are two possible ways to assign weights, viz., (i) Assign weights based on human interpretation, and (ii) Assign weights based on their chronological order. Assignment of weights based on human knowledge is not an acceptable solution for real world problems as human interpretation varies from one to another. Moreover, human interpretation is still an issue which is not understood by the computational scientists (Singh and Borah 2013a). Therefore, Yu (2005) considered the second way, where all the FLRs are given importance based on their chronological order. In this scheme, weight for each FLR is determined based on their sequence of occurrence.

That is, Yu (2005) gives more importance to events occurred recently than events occurred previously. However, this scheme of assigning weight is not justifiable, because it does not consider the severity of the events, which are represented by the fuzzy sets.

Problem 1.3.4 (*Utilization of first-order FLRs*). Most of the previous FTS models[4] use first-order FLRs (see Eq. 2.2.6) to get the forecasting results. The first-order FLRs based models use only immediately preceding value of the fuzzified time series for forecasting. Hence, the models which employ the first-order FLRs, are unable to capture more uncertainty that reside in the events.

Problem 1.3.5 (*Utilization of current state's fuzzy sets*). Previous FTS models (Song and Chissom 1993a, b, 1994) utilize the current state's fuzzified values in the right-hand side of the FLR, see Eq. 2.2.6) for forecasting. This approach, no doubt, improves the forecasting accuracy, but it degrades the predictive skill of the FTS models, because predicted values lie within the sample.

Problem 1.3.6 (*Calculation of defuzzified forecasted output*). In 1996, Chen (Chen 1996) used simplified arithmetic operations for calculation of forecasted output by avoiding the complicated max-min operations (see Eq. 2.3.2), and their method produced better results than Song and Chissom models.[5] Most of the existing FTS models[6] employ Chen's method (Chen 1996) to acquire the forecasting results. However, forecasting accuracy of these models are not good enough.

---

[4]References are: (Chen 1996; Cheng et al 2006; Hwang et al 1998; Song and Chissom 1993a, b, 1994).

[5]References are: (Song and Chissom 1993a, b, 1994).

[6]References are:(Egrioglu et al 2011; Huarng 2001b; Huarng and Yu 2006; Li and Chen 2004).

Problem 1.3.7 (*Advance prediction of ISMR*). The Indian economy is based on agriculture and its agri-products, and crop yield is heavily dependent on the summer monsoon (June–September) rainfall. Therefore, any decrease or increase in annual rainfall will always have a severe impact on the agricultural sector in India. About 65 % of the total cultivated land in India are under the influence of rain-fed agriculture system (Swaminathan 1998). Therefore, the prior knowledge of the monsoon behavior (during which the maximum rainfall occurs in a concentrated period) will help the Indian farmers and the Government to take advantage of the monsoon season. This knowledge can be very useful in reducing the damage to crops during less rainfall periods in the monsoon season. Therefore, forecasting the monsoon temporally is a major scientific issue in the field of monsoon meteorology.

The ensemble of statistics and mathematics has increased the accuracy of forecasting of ISMR up to some extent. But due to the non-linear nature of ISMR, its forecasting accuracy is still below the satisfactory level. In 2002, IMD failed to predict the deficit of rainfall during ISMR, which led to considerable concern in the meteorological community (Gadgil et al 2002). In 2004, again drought was observed in the country with a deficit of more than 13 % rainfall (Gadgil et al 2005), which could not be predicted by any statistical or dynamic model. Preethi et al (2011) reported that India as a whole received 77 % of rainfall during ISMR in 2009, which was the third highest deficient of all ISMR years during the period 1901–2009.

1.4 Research Contributions and Outline of the Book

This book uses SC techniques in order to deal with nonlinear and dynamic nature of the time series data. Several forecasting models are proposed considering the issues raised in the previous section. Contributions of this book along with the chapters' outlines are briefly explained below.

1. In Chap. 2, we introduce some fundamental concepts of the FTS modeling approach that will be widely adopted throughout the book. Along with that, we also present recent advancement followed by various SC techniques that are highly employed in the FTS modeling approach.
2. In Chap. 3, we present a new one-factor model to deal with four major problems of FTS forecasting approach, *viz.*, research Problems 1.3.1–1.3.3 and 1.3.6. To resolve the research Problem 1.3.1, a new "MBD" approach is proposed. To resolve the research Problem 1.3.2, we consider the approach proposed in Eq. 1.3.3. To resolve the research Problem 1.3.3, an index-based weight assignment technique is proposed. In this approach, weight for each FLR is determined by using index (i) of the fuzzy set (A_i) associated with the current state of the FLR. To explain this, consider the following FLRs as:

$$\begin{aligned} A_i &\rightarrow A_i \text{ with weight } i, \\ A_i &\rightarrow A_j \text{ with weight } j, \\ A_i &\rightarrow A_k \text{ with weight } k, \\ A_i &\rightarrow A_l \text{ with weight } l. \end{aligned} \quad (1.4.1)$$

In Eq. 1.4.1, each FLR is assigned a weight i, j, k, and l, based on indices of the current state of the FLRs, which are A_i, A_j, and A_k, and A_l, respectively. The advantage of using such approach is that the model can capture more persuasive weights of the FLRs based on the severity of the events during forecasting. To resolve the research Problem 1.3.6, we introduce a new "IBDT" in this chapter. Hence, the proposed model is entitled as "Efficient one-factor FTS forecasting model".

3. In Chap. 4, we present a new one-factor model based on hybridization of FTS theory with ANN. In this chapter, we deal with four major problems of FTS forecasting approach, *viz.*, research Problems 1.3.1, 1.3.4, 1.3.5 and 1.3.6. To resolve the research Problem 1.3.1, a new "RPD" approach is proposed. To resolve the research Problem 1.3.4, i.e., to capture more uncertainties of the events, we employ the high-order FLRs for forecasting. To resolve the research Problem 1.3.5, i.e., for obtaining the forecasting results out of sample (i.e., in advance), we use the previous state's fuzzified values (left hand side of FLRs, see Eq. 2.2.9) in this model. To defuzzify these fuzzified values, i.e., to resolve the research Problem 1.3.6, we develop an ANN based architecture, and incorporated it in this model. Hence, the proposed model is entitled as "High-order fuzzy-neuro time series forecasting model".
4. In Chap. 5, we present a new model to deal with the forecasting problems of two-factors. The proposed model is designed using hybridization of FTS with ANN. In this chapter, we deal with three major problems of FTS forecasting approach, *viz.*, research Problems 1.3.1, 1.3.4 and 1.3.6. To resolve the research Problem 1.3.1, i.e., for the creation of effective length of intervals of the historical time series data sets, an ANN based technique is adopted in this model. In this study, high-order FLRs (i.e., to resolve the research Problem 1.3.4) are also employed to design the model. To improve the efficiency of the model, we propose "Frequency-Weighing Defuzzication Technique" to defuzzify the fuzzified time series data sets (i.e., to resolve the research Problem 1.3.6). Hence, the proposed model is entitled as "Two-factors high-order neuro-fuzzy hybridized model".
5. In Chap. 6, we present a new model to deal with the forecasting problems of M-factors. The proposed model is designed using hybridization of Type-2 FTS with PSO. In this chapter, we deal with two major problems of FTS forecasting approach, viz., research Problems 1.3.1 and 1.3.6. For finding more effective lengths of intervals (i.e., to resolve the research Problem 1.3.1), recently many researchers applied the PSO algorithm on FTS models.[7] But, their applications

[7] References are: (Huang et al 2011; Kuo et al 2009, 2010; Park et al 2010; Sheikhan and Mohammadi 2012).

are limited to one-factor to two-factors time series data sets. In case of M-factors time series data set, large number of intervals are involved, which makes the determination of effective lengths of intervals very ineffective. These large number of intervals also affect the accuracy rate of forecasting. Motivated by this critical issue, we incorporate the PSO algorithm with the Type-2 FTS model. The main role of the PSO algorithm in the Type-2 FTS model is to improve the forecasting accuracy by adjusting the length of each interval in the universe of discourse and corresponding degree of membership simultaneously. To resolve the research Problem 1.3.6, we adopt the idea of defuzzification ("Frequency-Weighing Defuzzication Technique" as proposed in Chap. 5) for M-factors time series data set, and obtain the forecasting results. Hence, the proposed model is entitled as "M-factors based FTS-PSO model".

6. An attempt is made in Chap. 7 to resolve the problem of advanced prediction of ISMR (as discussed in Problem 1.3.7). For this purpose, an ANN based model is designed using the BBNN algorithm. Based on this algorithm, we have proposed five neural network architectures designated as BP1, BP2, ..., BP5 using three layers of neurons (one input layer, one hidden layer and one output layer). The detailed description of neural network architectures is also provided in this chapter. The proposed model predict ISMR of a given year using the observed time series data of the four monsoon months (June, July, August and September) and seasonal (sum of June, July, August and September).
7. Finally, the book conclusions and some future directions of work have been summarized in Chap. 8.

References

Chen SM (1996) Forecasting enrollments based on fuzzy time series. Fuzzy Sets Syst 81:311–319

Cheng C, Chang J, Yeh C (2006) Entropy-based and trapezoid fuzzification-based fuzzy time series approaches for forecasting IT project cost. Technol Forecast Social Change 73:524–542

Egrioglu E, Aladag CH, Yolcu U, Uslu V, Erilli N (2011) Fuzzy time series forecasting method based on Gustafson-Kessel fuzzy clustering. Expert Syst Appl 38(8):10355–10357

Gadgil S, Srinivasan J, Nanjundiah RS, Kumar KK, Munot AA, Kumar KR (2002) On forecasting the Indian summer monsoon: the intriguing season of 2002. Curr Sci 83(4):394–403

Gadgil S, Rajeevan M, Nanjundiah R (2005) Monsoon prediction-why yet another failure? Curr Sci 88(9):1389–1400

Gooijer JGD, Hyndman RJ (2006) 25 years of time series forecasting. Int J Forecast 22(3):443–473

Huang YL, Horng SJ, He M, Fan P, Kao TW, Khan MK, Lai JL, Kuo IH (2011) A hybrid forecasting model for enrollments based on aggregated fuzzy time series and particle swarm optimization. Expert Syst Appl 38(7):8014–8023

Huarng K (2001a) Effective lengths of intervals to improve forecasting in fuzzy time series. Fuzzy Sets Syst 123:387–394

Huarng K (2001b) Heuristic models of fuzzy time series for forecasting. Fuzzy Sets Syst 123:369–386

Huarng K, Yu THK (2006) Ratio-based lengths of intervals to improve fuzzy time series forecasting. IEEE Trans Syst Man Cybern Part B Cybern 36(2):328–340

Huarng KH, Yu THK, Hsu YW (2007) A multivariate heuristic model for fuzzy time-series forecasting. IEEE Trans Syst Man Cybern Part B: Cybern 37:836–846

Hwang JR, Chen SM, Lee CH (1998) Handling forecasting problems using fuzzy time series. Fuzzy Sets Syst 100:217–228

Kuo IH, Horng SJ, Kao TW, Lin TL, Lee CL, Pan Y (2009) An improved method for forecasting enrollments based on fuzzy time series and particle swarm optimization. Expert Syst Appl 36(3, Part 2):6108–6117

Kuo IH, Horng SJ, Chen YH, Run RS, Kao TW, Chen RJ, Lai JL, Lin TL (2010) Forecasting TAIFEX based on fuzzy time series and particle swarm optimization. Expert Syst Appl 37(2):1494–1502

Li ST, Chen YP (2004) Natural partitioning-based forecasting model for fuzzy time-series. IEEE Int Conf Fuzzy Syst 3:1355–1359

Mahnam M, Ghomi SMTF (2012) A particle swarm optimization algorithm for forecasting based on time variant fuzzy time series. Int J Ind Eng Prod Res 23(4):269–276

Park J, Lee DJ, Song CK, Chun MG (2010) TAIFEX and KOSPI 200 forecasting based on two-factors high-order fuzzy time series and particle swarm optimization. Expert Syst Appl 37(2):959–967

Preethi B, Revadekar JV, Kripalani RH (2011) Anomalous behaviour of the indian summer monsoon 2009. J Earth Syst Sci 120(5):783–794

Sheikhan M, Mohammadi N (2012) Time series prediction using PSO-optimized neural network and hybrid feature selection algorithm for IEEE load data. Neural Comput Appl

Singh P, Borah B (2012) An effective neural network and fuzzy time series-based hybridized model to handle forecasting problems of two factors. Knowl Inf Syst 38(3):669–690

Singh P, Borah B (2013a) An efficient time series forecasting model based on fuzzy time series. Eng Appl Artif Intell 26:2443–2457

Singh P, Borah B (2013b) Forecasting stock index price based on M-factors fuzzy time series and particle swarm optimization. Int J Approx Reason

Singh P, Borah B (2013c) High-order fuzzy-neuro expert system for daily temperature forecasting. Knowl-Based Syst 46:12–21

Singh P, Borah B (2013d) Indian summer monsoon rainfall prediction using artificial neural network. Stoch Environ Res Risk Assess 27(7):1585–1599

Song Q, Chissom BS (1993a) Forecasting enrollments with fuzzy time series—Part I. Fuzzy Sets Syst 54(1):1–9

Song Q, Chissom BS (1993b) Fuzzy time series and its models. Fuzzy Sets Syst 54(1):1–9

Song Q, Chissom BS (1994) Forecasting enrollments with fuzzy time series - Part II. Fuzzy Sets Syst 62(1):1–8

Swaminathan MS (1998) Padma Bhusan Prof. P. Koteswaram First memorial lecture-23rd March 1998. In: Climate and sustainable food security, vol 28, Vayu Mandal, pp 3–10

Yu HK (2005) Weighted fuzzy time series models for TAIEX forecasting. Phys A: Stat Mech Appl 349(3–4):609–624

Chapter 2
Fuzzy Time Series Modeling Approaches: A Review

Although this may seem a paradox, all exact science is dominated by the concept of approximation.
By Bertrand Shaw (1872–1970)

Abstract Recently, there seems to be increased interest in time series forecasting using soft computing (SC) techniques, such as fuzzy sets, artificial neural networks (ANNs), rough set (RS) and evolutionary computing (EC). Among them, fuzzy set is widely used technique in this domain, which is referred to as "Fuzzy Time Series (FTS)". In this chapter, extensive information and knowledge are provided for the FTS concepts and their applications in time series forecasting. This chapter reviews and summarizes previous research works in the FTS modeling approach from the period 1993–2013 (June). Here, we also provide a brief introduction to SC techniques, because in many cases problems can be solved most effectively by integrating these techniques into different phases of the FTS modeling approach. Hence, several techniques that are hybridized with the FTS modeling approach are discussed briefly. We also identified various domains specific problems and research trends, and try to categorize them. The chapter ends with the implication for future works. This review may serve as a stepping stone for the amateurs and advanced researchers in this domain.

Keywords Fuzzy time series (FTS) · Artificial neural networks (ANNs) · Rough set (RS) · Evolutionary computing (EC)

2.1 Soft Computing: An Introduction

The term "soft computing" is a multidisciplinary field which pervades from a mathematical science to computer science, information technology, engineering applications, etc. The conventional computing or hard-computing generally deals with precision, certainty and rigor (Zadeh 1994). However, the main desiderata of SC is

P. Singh, *Applications of Soft Computing in Time Series Forecasting*,
Studies in Fuzziness and Soft Computing 330, DOI 10.1007/978-3-319-26293-2_2

to tolerate with imprecision, uncertainty, partial truth, and approximation (Yardimci 2009). SC is influenced by many researchers. Among them, Zadeh's contribution is invaluable. Zadeh published his most influential work in SC in 1965 (Zadeh 1965). Later, he contributed in this area by publishing numerous research articles on the analysis of complex systems and decision processes (Zadeh 1973), approximate reasoning (Zadeh 1975a, b), knowledge representation (Zadeh 1989), design and deployment of intelligent systems (Zadeh 1997), etc. According to Jang et al. (1997), "SC is not a single methodology. Rather, it is a partnership in which each of the partners contributes a distinct methodology for addressing problems in its domain." Therefore, SC has been evolving as an amalgamated field of different methodologies such as *fuzzy sets, ANN, EC and probabilistic computing* (Dote and Ovaska 2001; Herrera-Viedma et al. 2014; Kacprzyk 2010; Szmidt et al. 2014). Later, *RS, chaos computing and immune network theory* have been included into SC (Castro and Timmis 2003; Mitra et al. 2002). The main objective of hybridizing these methodologies is to design an intelligent machine and find solution to nonlinear problems which can not be modeled mathematically (Zadeh 2002).

2.1.1 Time Series Events and Uncertainty

A time series represents a collection of values of certain events or tasks which are obtained with respect to time. Advance prediction of some significant time series events such as temperature, rainfall, stock price, population growth, economic growth, etc., are major scientific issues in the domain of forecasting. Imprecise knowledge or information cannot be overlooked in this domain. Because of the nature of the time series data, which is highly non-stationary and uncertain, the decision-making process becomes very tedious. For example, sudden rise and fall of daily temperature, sudden increase and decrease of daily stock index price, sudden increase and decrease of rainfall amount indicate that these events are very uncertain. The characteristics of all these events cannot be described accurately; therefore, it is referred to as "imprecise knowledge" or "incomplete knowledge". Due to these problems, mathematical or statistical models can not deal with this imprecise knowledge, thereby diluting the accuracy very significantly.

Future prediction of time series events has attracted people from the beginning of times. However, forecasting these events with 100 % accuracy may not be possible, their forecasting accuracy and the speed of forecasting process can be improved. To resolve this problem, Song and Chissom (1993a) developed a model in 1993 based on uncertainty and imprecise knowledge contained in time series data. They initially used the fuzzy sets concept to represent or manage all these uncertainties, and referred this concept as "Fuzzy Time Series (FTS)".

Forecasting the short term time series events are frequently attempted by the researchers, and its accuracy is better than long term predictions. From 1994 onwards, researchers have developed numerous models based on the FTS concept to deal with the forecasting problems of short term as well as long term events. This study focuses on the application and use of fuzzy sets concept in forecasting of such events. The basic knowledge of ANNs, RS and EC are provided complimentary with the sound background of fuzzy sets, because in many cases a problem can be solved most effectively by hybridizing these techniques together rather independently. Hence, one of the objectives of this chapter is also to introduce the SC methodologies (such as ANNs, RS and EC) that are employed by the FTS modeling approach to represent and manage the imprecise knowledge in time series forecasting.

2.2 Definitions

In this section, we provide various definitions for the terminologies used throughout this book.

Definition 2.2.1 (*Time Series*) (Brockwell and Davis 2008). A time series is a set of observations x_t, each one being recorded at specific time t. When observations are made at fixed time intervals, then it is called a "discrete-time series". If observations are recorded continuously over some time interval, then is called a "continuous-time series".

The main objective of time series forecasting is reckoning the future values of the series. In literature, several time series forecasting models are available (Chatfield 2000). Forecasting model finds optimal forecasts based on the type of data and condition of the model. Suppose we have an observed time series $x_1, x_2, \ldots, x_N$ and wish to forecast future values such as x_{N+h}. Forecasting model can make the lead time forecasts (denoted as h), or make forecast h steps ahead of time N.

This survey is concerned with the study of SC techniques and its application in FTS modeling approach. Therefore, in the next, we will discuss initially the fuzzy sets concept and its application in time series forecasting.

In 1965, Zadeh (1965) introduced fuzzy sets theory involving continuous set membership for processing data in presence of uncertainty. He also presented fuzzy arithmetic theory and its application in these articles.[1]

Definition 2.2.2 (*Universe of discourse*) (Song and Chissom 1993a). Let L_{bd} and U_{bd} be the lower bound and upper bound of the time series data, respectively. Based on L_{bd} and U_{bd}, we can define the universe of discourse U as:

$$U = [L_{bd}, U_{bd}] \tag{2.2.1}$$

Definition 2.2.3 (*Fuzzy Set*) (Zadeh 1965). A fuzzy set is a class with varying degrees of membership in the set. Let U be the universe of discourse, which is discrete and finite, then fuzzy set A can be defined as follows:

---

[1]References are: (Zadeh 1971, 1973, 1975a).

$$A = \{\mu_{A(x_1)}/x_1 + \mu_{A(x_2)}/x_2 + \ldots\} = \Sigma_i \mu_A(x_i)/x_i \tag{2.2.2}$$

where μ_A is the membership function of A, $\mu_A: U \to [0, 1]$, and $\mu_{A(x_i)}$ is the degree of membership of the element x_i in the fuzzy set A. Here, the symbol "+" indicates the operation of union and the symbol "/" indicates the separator rather than the commonly used summation and division in algebra, respectively.

When U is continuous and infinite, then the fuzzy set A of U can be defined as:

$$A = \{\int \mu_{A(x_i)}/x_i\}, \forall x_i \in U \tag{2.2.3}$$

where the integral sign stands for the union of the fuzzy singletons, $\mu_{A(x_i)}/x_i$.

Definition 2.2.4 (*FTS*).[2] Let $Y(t)(t = 0, 1, 2, \ldots)$ be a subset of Z and the universe of discourse on which fuzzy sets $\mu_i(t)(i = 1, 2, \ldots)$ are defined and let $F(t)$ be a collection of $\mu_i(t)(i = 1, 2, \ldots)$. Then, $F(t)$ is called a FTS on $Y(t)(t = 0, 1, 2, \ldots)$.

With the help of the following two examples, the notions of FTS can be explained:
[Example 1] The common observations of daily weather condition for certain region can be described using the daily common words "hot", "very hot", "cold", "very cold", "good", "very good", etc. All these words can be represented by fuzzy sets.
[Example 2] The common observations of the performance of a student during the final year of degree examination can be represented using the fuzzy sets "good", "very good", "poor", "bad", "very bad", etc.

These two examples represent the processes, and conventional time series models are not applicable to describe these processes (Song and Chissom 1993b). Therefore, Song and Chissom (Song and Chissom 1993b) first time used the fuzzy sets notion in time series forecasting. Later, their proposed method has gained in popularity in scientific community as a "FTS forecasting model".

Definition 2.2.5 (*Fuzzification*) (Zadeh 1975a). The operation of *fuzzification* transforms a nonfuzzy set (crisp set) into a fuzzy set or increasing the fuzziness of a fuzzy set. Thus, a *fuzzifier* R is applied to a fuzzy subset i of the universe of discourse U yields a fuzzy subset $R(i; T)$, which can be expressed as:

$$R(i; T) = \int_U \mu_i(u)T(u), \tag{2.2.4}$$

where the fuzzy set $T(u)$ is the kernel of R, i.e., the result of applying R to a singleton $1/u$:

$$T(u) = R(1/u; T) \tag{2.2.5}$$

where $\mu_i(u)T(u)$ represents the product of a scalar $\mu_i(u)$ and the fuzzy set $T(u)$; and $\int$ is the union of the family of fuzzy sets $\mu_i(u)T(u)$, $u \in U$.

[2]References are: (Song and Chissom 1993a, b, 1994).

Definition 2.2.6 (*Defuzzification*) (Ross 2007). Defuzzification of a fuzzy set is the process of "rounding it off" from its location in the unit hypercube to the nearest vertex, i.e., it is the process of converting a fuzzy set into a crisp set.

Definition 2.2.7 (*FLR*).[3] Assume that $F(t-1) = A_i$ and $F(t) = A_j$. The relationship between $F(t)$ and $F(t-1)$ is referred to as a FLR, which can be represented as:

$$A_i \rightarrow A_j, \tag{2.2.6}$$

where A_i and A_j refer to the left-hand side and right-hand side of the FLR, respectively.

Definition 2.2.8 (*FLRG*).[4] Assume the following FLRs as follows:

$$\begin{aligned} &A_i \rightarrow A_{k1}, \\ &A_i \rightarrow A_{k2}, \\ &A_i \rightarrow A_{k3}, \\ &A_i \rightarrow A_{k4}, \\ &\quad\dots \\ &A_i \rightarrow A_{km} \end{aligned} \tag{2.2.7}$$

Chen (1996) suggested that FLRs having the same fuzzy sets on the left-hand side can be grouped into a FLRG. So, based on Chen's model (Chen 1996), these FLRs can be grouped into the FLRG as:

$$A_i \rightarrow A_{k1}, A_{k2}, A_{k3}, A_{k4} \dots, A_{km}. \tag{2.2.8}$$

Definition 2.2.9 (*High-order FLR*) (Chen and Chen 2011a). Assume that $F(t)$ is caused by $F(t-1)$, $F(t-2)$, ..., and $F(t-n)$ $(n > 0)$, then high-order FLR can be expressed as:

$$F(t-n), \dots, F(t-2), F(t-1) \rightarrow F(t) \tag{2.2.9}$$

Definition 2.2.10 (*M-factors FTS*). Let FTS $A(t), B(t), C(t), \dots, M(t)$ be the factors/observations of the forecasting problems. If we only use $A(t)$ to solve the forecasting problems, then it is called a one-factor FTS. If we use remaining secondary-factors/secondary-observations $B(t), C(t), \dots, M(t)$ with $A(t)$ to solve the forecasting problems, then it is called M-factors FTS.

[3]References are: (Chen 1996; Song and Chissom 1993a, 1994).

[4]References are: (Chen 1996; Song and Chissom 1993a, 1994).

One-factor FTS models (referred to as Type-1 FTS models) employ only one variable for forecasting (Hsu et al. 2003; Huarng 2001). For example, researchers in these articles (Chen et al 2008; Hsu et al. 2003) consider only *closing* price in the forecasting of the stock index. However, the stock index price consists of many different observations, such as *opening*, *high*, *low*, etc. If these additional observations are used with one-factor variable, then it is referred to as M-factors FTS model. The model proposed by Huarng and Yu (Huarng and Yu 2005) is based on M-factors, because they use *high* and *low* as the secondary-observations to forecast the *closing* price of TAIEX.

Definition 2.2.11 (*Type-2 fuzzy set*) (Greenfield and Chiclana 2013). Let $A(U)$ be the set of fuzzy sets in U. A Type-2 fuzzy set A in X is fuzzy set whose membership grades are themselves fuzzy. This implies that $\mu_A(x)$ is a fuzzy set in U for all x, i.e., $\mu_A : X \rightarrow A(U)$ and

$$A = \{(x, \mu_A(x)) | \mu_A(x) \in A(U) \forall x \in X\} \tag{2.2.10}$$

The concept of Type-2 fuzzy set is explained with an example as follows:

[Explanation] When we cannot distinguish the degree of membership of an element in a set as 0 or 1, we use Type-1 fuzzy sets. Similarly, when the nature of an event is so fuzzy so that determination of degree of membership as a crisp number in the range [0, 1] is so difficult, then we use Type-2 fuzzy sets (Mencattini et al. 2005). This Type-2 fuzzy sets concept was first introduced by Zadeh (1975a) in 1975. In Type-1 fuzzy set, the degree of membership is characterized by a crisp value; whereas in Type-2 fuzzy set, the degree of membership is regarded as a fuzzy set (Chen 2012). Thus, if there are more uncertainty in the event, and we have difficulty in determining its exact value, then we simply use Type-1 fuzzy sets, rather than crisp sets. But, ideally we have to use some finite-type sets, just like Type-2 fuzzy sets (Mencattini et al. 2005). Based on this explanation, we present an example which is based on article Huarng and Yu (2005) as follows:

Let us consider a fuzzy set for "Closing Price" of stock index, as shown in Fig. 2.1 (left). Here, we have a crisp degree of membership values 1.0 and 0.5 for the "Closing Price = 1000" and "Closing Price = 500", respectively. Based on the above explanation, the "Closing Price = 1000" can have more than one degree of memberships. For example, in Fig. 2.1 (right), there are three degrees of memberships (0.4, 0.5 and 0.6) for the "Closing Price = 1000". In other words, there can be multiple degrees of membership for the same "Closing Price = 1000", as shown in Fig. 2.1 (right). In Fig. 2.1 (right), the highest degree of membership (0.6) indicates the positive view about the occurrence of event, whereas the lowest degree of membership (0.4) indicates the negative view about the occurrence of event. We can use these positive

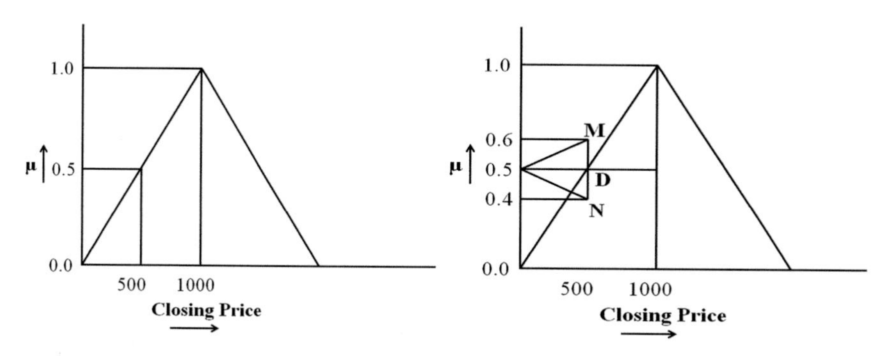

Fig. 2.1 A Type-1 (*left*) and Type-2 (*right*) fuzzy sets

and negative views together in FTS modeling approach. In summary, we can use more observations/information from the positive and negative views for forecasting in each time period.

Definition 2.2.12 (*Type-2 FTS model*) (Huarng and Yu 2005). A Type-2 FTS model can be defined as an extension of a Type-1 FTS model. The Type-2 FTS model employs the FLRs established by a Type-1 model based on Type-1 observations. Fuzzy operators such as union and intersection are used to establish the new FLRs obtained from Type-1 and Type-2 observations. Then, Type-2 forecasts are obtained from these FLRs.

2.3 FTS Modeling Approach

Chen (1996) proposed a simple calculation method to get a higher forecasting accuracy in FTS model. Still this model is used as the basis of FTS modeling. The basic architecture of this model is depicted in Fig. 2.2. This model employs the following five common steps to deal with the forecasting problems of time series, which are explained below. Contributions of various research articles in different phases of this model are also categorized in this section.

step 1. *Partition the universe of discourse into intervals.* The universe of discourse can be defined based on Eq. 2.2.1. After determination of length of intervals, U can be partitioned into several equal lengths of intervals. For determining the universe of discourse and to partition them into effective lengths of intervals, many researchers provide various solutions in these articles (Chen and Tanuwijaya 2011; Cheng et al. 2008a; Lee and Chou 2004; Li and Cheng 2007; Liu and Wei 2010; Yu 2005a). Some recent advancement in this step can be found in these articles (Bang and Lee 2011; Chen and Wang 2010; Cheng et al. 2011; Egrioglu et al. 2010, 2011a, b; Gangwar and Kumar

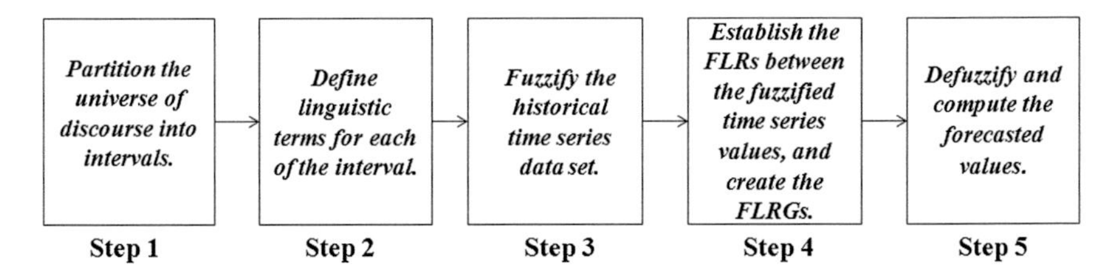

Fig. 2.2 Architecture of Chen's Model

2012; Huarng and Yu 2006b; Li et al. 2010, 2011; Liu et al. 2009; Wong et al. 2010).

step 2. *Define linguistic terms for each of the interval.* After generating the intervals, linguistic terms are defined for each of the interval. In this step, we assume that the historical time series data set is distributed among n intervals (i.e., $a_1, a_2, \ldots,$ and a_n). Then, define n linguistic variables $A_1, A_2, \ldots, A_n$, which can be represented by fuzzy sets, as shown below:

$$\begin{aligned}
A_1 &= 1/a_1 + 0.5/a_2 + 0/a_3 + \ldots + 0/a_{n-2} + 0/a_{n-1} + 0/a_n,\\
A_2 &= 0.5/a_1 + 1/a_2 + 0.5/a_3 + \ldots + 0/a_{n-2} + 0/a_{n-1} + 0/a_n,\\
A_3 &= 0/a_1 + 0.5/a_2 + 1/a_3 + \ldots + 0/a_{n-2} + 0/a_{n-1} + 0/a_n,\\
&\vdots\\
A_n &= 0/a_n + 0/a_2 + 0/a_3 + \ldots + 0/a_{n-2} + 0.5/a_{n-1} + 1/a_n.
\end{aligned} \tag{2.3.1}$$

Then, we obtain the degree of membership of each time series value belonging to each A_i. Here, maximum degree of membership of fuzzy set A_i occurs at interval a_i, and $1 \leq i \leq n$. Then, each historical time series value is fuzzified. For example, if any time series value belongs to the interval a_i, then it is fuzzified into A_i, where $1 \leq i \leq n$.

For ease of computation, the degree of membership values of fuzzy set $A_j(j = 1, 2, \ldots, n)$ are considered as either 0, 0.5 or 1, and $1 \leq i \leq n$. In Eq. 2.3.1, for example, A_1 represents a linguistic value, which denotes a fuzzy set $= \{a_1, a_2, \ldots, a_n\}$. This fuzzy set consists of n members with different degree of membership values $= \{1, 0.5, 0, \ldots, 0\}$. Similarly, the linguistic value A_2 denotes the fuzzy set $= \{a_1, a_2, \ldots, a_n\}$, which also consists of n members with different degree of membership values $= \{0.5, 1, 0.5, \ldots, 0\}$. The descriptions of remaining linguistic variables, *viz.*, $A_3, A_4, \ldots, A_n$, can be provided in a similar manner.

Since each fuzzy set contains n intervals, and each interval corresponds to all fuzzy sets with different degree of membership values. For example, interval a_1 corresponds to linguistic variables A_1 and A_2 with degree of membership values 1 and 0.5, respectively, and remaining fuzzy sets with degree of membership value 0. Similarly, interval a_2 corresponds to linguistic variables A_1, A_2 and A_3 with degree of membership values 0.5, 1, and 0.5, respectively, and remaining fuzzy sets with degree of membership value 0. The descriptions of remaining intervals, *viz.*, $a_3, a_4, \ldots, a_n$, can be provided in a similar manner.

Liu (2007) introduced an improved FTS forecasting method in which the forecasted value is regarded as a trapezoidal fuzzy number instead of a single-point value. They replace the above discrete fuzzy sets (as discussed in Eq. 2.3.1) with trapezoidal fuzzy numbers. The main advantage of the proposed method is that the decision analyst can accumulate information about the possible forecasted ranges under different degrees of confidence.

step 3. *Fuzzify the historical time series data set.* In order to fuzzify the historical time series data, it is essential to obtain the degree of membership value of each observation belonging to each A_j $(j = 1, 2, \ldots, n)$ for each day/year. If the maximum membership value of one day's/year's observation occurs at interval a_i and $1 \leq i \leq n$, then the fuzzified value for that particular day/year is considered as A_i.

In FTS model, each fuzzy set carries the information of occurrence of the historic event in the past. So, if these fuzzy sets would not be handled efficiently, then important information may be lost. Therefore, for fuzzification purpose, many researchers provided different techniques in these articles (Cheng et al. 2006; Hwang et al. 1998; Sah and Degtiarev 2005).

step 4. *Establish the FLRs between the fuzzified time series values, and create the FLRGs.* After time series data is completely fuzzified, then FLRs have been established based on Definition 2.2.7. The first-order FLR is established based on two consecutive linguistic values. For example, if the fuzzified values of time $t - 1$ and t are A_i and A_j, respectively, then establish the first-order FLR as "$A_i \rightarrow A_j$", where "A_i" and "A_j" are called the previous state and current state of the FLR, respectively. Similarly, the nth-order FLR is established based on $n + 1$ consecutive linguistic values. For example, if the fuzzified values of time $t - 4$, $t - 3$, $t - 2$, $t - 1$ and t are A_{ai}, A_{bi}, A_{ci}, A_{di} and A_{ej}, respectively, then the fourth-order FLR can be established as "$A_{ai}, A_{bi}, A_{ci}, A_{di} \rightarrow A_{ej}$", where "$A_{ai}, A_{bi}, A_{ci}, A_{di}$" and "$A_{ej}$" are called the previous state nd current state of the FLR, respectively.

Most of the existing FTS models[5] use the first-order FLRs to get the forecasting results. In these articles,[6] researchers show that the high-order FLRs (see Definition 2.2.9) can improve the forecasting accuracy. The main reason of obtaining high accuracy from these high-order FTS models is that it can consider more linguistic values that represent the high uncertainty

[5]References are: (Chang et al. 2007; Chen 1996; Cheng et al. 2006; Huarng 2001; Hwang et al. 1998; Song and Chissom 1993a, b, 1994).

[6]References are: (Aladag et al. 2009, 2010; Avazbeigi et al. 2010; Bahrepour et al. 2011; Chen 2002; Chen and Chen 2011a, b; Chen and Chung 2006b; Chen et al 2008; Gangwar and Kumar 2012; Jilani and Burney 2008; Own and Yu 2005; Singh 2007a, c, 2008, 2009; Tsai and Wu 2000).

involved in various dynamic processes. On the other hand, to extract rule from the fuzzified time series data set, Qiu et al. (2012) utilized C-fuzzy decision trees (Pedrycz and Sosnowski 2005) in FTS model. They introduced two major improvements in C-fuzzy decision trees, *viz.*, first a new stop condition is introduced to reduce the computational cost, and second weighted C-fuzzy decision tree (WCDT) is introduced where weight distance is computed with information gain. In this approach, the forecast rule are expressed as "if input value is ... then it can be labeled as ...".
Based on the same previous state of the FLRs, the FLRs can be grouped into a FLRG (see Definition 2.2.8). For example, the FLRG "$A_i \rightarrow A_m, A_n$" indicates that there are following FLRs:

$$A_i \rightarrow A_m,$$
$$A_i \rightarrow A_n.$$

Step 5. *Defuzzify and compute the forecasted values.* In these articles (Song and Chissom 1993a; Tsaur et al. 2005), researchers adopted the following method to forecast enrollments of the University of Alabama:

$$Y(t) = Y(t-1) \circ R, \tag{2.3.2}$$

where $Y(t-1)$ is the fuzzified enrollment of year $(t-1)$, $Y(t)$ is the forecasted enrollment of year t represented by fuzzy set, "$\circ$" is the max-min composition operator, and "R" is the union of fuzzy relations. This method takes much time to compute the union of fuzzy relations R, especially when the number of fuzzy relations is more in Eq. 2.3.2 (Chen and Hwang 2000; Huarng et al 2007). Therefore, some researchers in these articles[7] introduced various solutions for the defuzzification operation. One of the solution introduced by Chen (Chen 1996) is presented below.
This includes the following two principles, *viz.*, **Principle 1** and **Principle 2**. The procedure for **Principle 1** is given as follows:

- **Principle 1**: For forecasting $F(t)$, the fuzzified value for $F(t-1)$ is required, where "t" is the current time which we want to forecast. The **Principle 1** is applicable only if there are more than one fuzzified values available in the current state. The steps under **Principle 1** are explained next.

Step 1. Obtain the fuzzified value for $F(t-1)$ as A_i $(i = 1, 2, 3 \ldots, n)$.
Step 2. Obtain the FLR whose previous state is A_i and the current state is $A_{j1}, A_{j2}, \ldots, A_{jp}$, i.e., the FLR is in the form of "$A_i \rightarrow A_{j1}, A_{j2}, \ldots, A_{jp}$".

[7] References are: (Chen 1996, 2002; Cheng et al. 2008b; Huarng 2001; Huarng et al 2007; Hwang et al. 1998; Jilani and Burney 2008; Kuo et al. 2009; Lee et al. 2006; Li et al. 2008; Qiu et al 2011; Singh and Borah 2012, 2013b; Singh 2007a, b, 2009; Yu 2005b).

Step 3. Find the interval where the maximum membership value of the fuzzy sets $A_{j1}, A_{j2}, \ldots, A_{jp}$ (current state) occur, and let these intervals be $a_{j1}, a_{j2}, \ldots, a_{jp}$. All these intervals have the corresponding mid-values $C_{j1}, C_{j2}, \ldots, C_{jp}$.

Step 3. Compute the forecasted value as:

$$Forecasted_{value} = \left[\frac{C_{j1} + C_{j2} + \ldots + C_{jp}}{p}\right] \tag{2.3.3}$$

Here, p represents the total number of fuzzy sets associated with the current state of the FLR.

- **Principle 2**: This principle is applicable only if there is only one fuzzified value in the current state. The steps under **Principle 2** are given as follows:

Step 1. Obtain the fuzzified value for $F(t-1)$ as A_i $(i = 1, 2, \ldots, n)$.

Step 2. Find the FLR whose previous state is A_i and the current state is A_j, i.e., the FLR is in the form of "$A_i \rightarrow A_j$".

Step 3. Find the interval where the maximum membership value of the fuzzy set A_j occurs. Let these interval be a_j $(j = 1, 2, 3, \ldots, n)$. This interval a_j has the corresponding mid-value C_j. This C_j is the forecasted value for $F(t)$.

2.4 Hybridize Modeling Approach for FTS

Recently, several SC techniques have been employed to deal with the different challenges imposed by the FTS modeling approach. The main SC techniques for this purpose include ANN, RS, and EC. Each of them provides significant solution for addressing domain specific problems. The combination of these techniques leads to the development of new architecture, which is more advantageous and the expert, providing robust, cost effective and approximate solution, in comparison to conventional techniques. However, this hybridization should be carried out in a reasonable, rather than an expensive or a complicated, manner.

In the following, we describe the basics of individual SC techniques and their hybridization techniques, along with the several hybridized models developed for handling forecasting problems of the FTS modeling approach. It should be noted that still there is no any universally recognized method to select particular SC technique(s), which is suitable for resolving the problems. The selection of technique(s) is completely dependent on the problem and its application, and requires human interpretation for determining the suitability of a particular technique.

2.4.1 ANN: An Introduction

ANNs are massively parallel adaptive networks of simple nonlinear computing elements called *neurons* which are intended to abstract and model some of the functionality of the human nervous system in an attempt to partially capture some of its computational strengths (Kumar 2004). The neurons in an ANN are organized into different layers. Inputs to the network are entered in the input layer; whereas outputs are produced as signals in the output layer. These signals may pass through one or more intermediate or *hidden* layers which transform the signals depending upon the neuron signal functions.

The neural networks are classified into either single-layer or multi-layer. In multi-layer networks hidden layers exist between input layer and output layer. A single-layer feed-forward (SLFF) neural network is formed when the nodes of input layer are connected with output nodes with various weights. A multi-layer feed-forward (MLFF) neural network architecture can be developed by increasing the number of layers in SLFF neural network. Feed-forward ANNs allow signals to travel from input to output. There is no feed-back loop. Feed-back networks can have signals travelling in both directions by introducing loops in the network. Feed-back networks are also referred to as interactive or recurrent networks.

Usually FFNN are used in time series forecasting. Recurrent networks are also used in some cases. Researchers employ ANN in various forecasting problems such as electric load forecasting (Taylor and Buizza 2002), short-term precipitation forecasting (Kuligowski and Barros 1998), credit ratings forecasting (Kumar and Bhattacharya 2006), tourism demand forecasting (Law 2000) etc., due to its capability to discover complex nonlinear relationships (Czibula et al. 2013; Donaldson and Kamstra 1996; Indro et al. 1999) in the observations. More detailed description on applications of ANN (especially BPNN) can be found in article written by Wilson et al. (2002).

Multi-layer FFNN uses back-propagation learning algorithm, therefore such networks are also known as back-propagation networks (BPNN). The main objective of using BPNN is to minimize the output error obtained from the difference between the calculated output $(o_1, o_2, \ldots, o_n)$ and target output $(n_1, n_2, \ldots, n_n)$ of the neural network by adjusting the weights (see Fig. 2.3). So in BPNN, each information is sent back again in the reverse direction until the output error is very small or zero. BPNN is trained under the process of three phases: (a) feed-forward of the input training pattern, (b) the calculation and back-propagation of the associated error, and (c) the adjustment of the weights.

Due to large number of additional parameters (e.g., initial weight, learning rate, momentum, epoch, activation function, etc.), an ANN model has great capability to learn by making proper adjustment of these parameters, in order to produce the desired output. During the training process, this output may fit the data very well, but it may produce poor results during the testing process. This implies that the neural network may not generalize well. This might be caused due to *overfitting* or *overtraining* of data (Weigend 1994), which can be controlled by monitoring the error

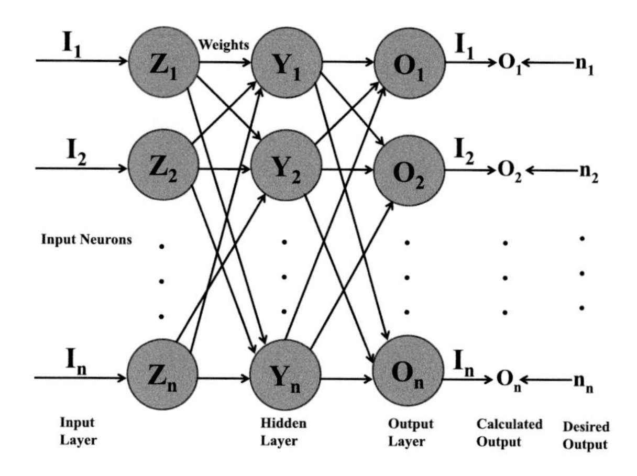

Fig. 2.3 A BPNN architecture with one hidden layer

during training process and terminate the process when the error reaches a minimum threshold with respect to the testing set (Sivanandam and Deepa 2007; Venkatesan et al. 1997). Another way to make the neural network generalize enough so that it performs well for training and testing data is to make small changes in the number of layers and neurons in the input space, without changing the output components. However, choosing the best neural network architecture is a heuristic approach. One solution to this problem is to keep the architecture of neural network relatively simple and small (Beale et al. 2010), because complex architectures are much more prone to overfitting (Gaume and Gosset 2003; Haykin 1999; Liu et al. 2008; Piotrowski and Napiorkowski 2012). Therefore, Hornik et al. (1989) suggested to design the neural network architecture with optimum number of neurons in the single hidden layer. Tan et al. (2009) also stated that one of the best ways to do this is to construct a fully connected neural network with a sufficiently large number of neurons in the hidden layer, and then iterate the architecture-building process with a smaller number of neurons.

Hybridization of ANN with FTS is a significant development in the domain of forecasting. It is an ensemble of the merits of ANN and FTS, by substituting the demerits of one technique by the merits of another technique. This includes various advantages of ANN, such as parallel processing, handling of large data set, fast learning capability, etc. Handling of imprecise/ uncertain and linguistic variables are done through the utilization of fuzzy sets. Besides these advantages, the FTS-ANN hybridization help in designing complex decision-making systems.

ANN can be used in different steps of FTS modeling approach. These steps are discussed in Sect. 2.3. In Fig. 2.4, three different hybridized architectures are presented, where applications of ANN are demonstrated in different steps of FTS modeling approach. In the first architecture, ANN is responsible for determination of FLRs (top); in the second architecture, ANN is responsible for partitioning the universe of discourse (middle); and in the third architecture, ANN is responsible

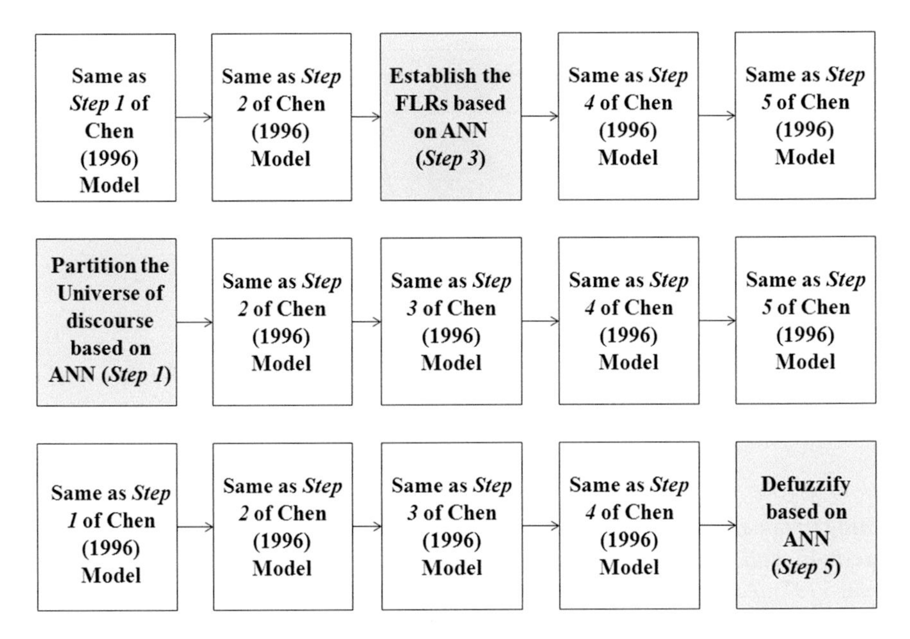

Fig. 2.4 Block diagrams of FTS-ANN hybridized models

for defuzzification operation (bottom). The roles of ANN in these architectures are explained below.

(a) *For defining FLRs*: In this case, primary inputs for connection-oriented neural network are fuzzified time series values. The neural network is trained in terms of the number of input nodes, hidden nodes and desired outputs. One or more hidden layers are employed to automatically generate the FLRs, which may later be clustered into similar FLRGs.
In the articles (Aladag et al. 2009, 2010), researchers employ FFNN to define high-order FLRs in FTS model. Both these models are applied in forecasting the enrollments of the University of Alabama. Similar to these two approaches, many researchers (Egrioglu et al. 2009; Huarng and Yu 2006a, 2012; Yolcu et al 2011; Yu and Huarng 2010) use the ANN in FTS model to capture the FLRs for improving the forecasted accuracy.
For defining high-order FLRs, a neural network architecture for the nth-order FLRs is shown in Fig. 2.5. Here, each input node take the previous days $F(t - n), \ldots, F(t - 2), F(t - 1)$ fuzzified time series values, e.g., $A_l, \ldots, A_m, A_n$ respectively to predict current day $F(t)$ fuzzified time series value, e.g., A_j. Here, each "t" represents the day for corresponding fuzzified time series values. Based on the input and output fuzzified values, the nth-order FLRs are established as: $A_l, \ldots, A_m, A_n \rightarrow A_j$. During simulation, the indices of previous state fuzzy sets (e.g., $l, \ldots, m, n$) are used as inputs, whereas index of current state fuzzy set (e.g., j) is used as target output.

(b) *For partitioning the Universe of discourse*: Data clustering is a popular approach for automatically finding classes, concepts, or groups of patterns (Gondek and Hofmann 2007). Time series data are pervasive across all human endeavors, and their clustering is one of the most fundamental applications of data mining (Keogh and Lin 2005). In literature, many data clustering algorithms (Estivill-Castro 2002; Ordonez 2003; Wu et al. 2008) have been proposed, but their applications are limited to the extraction of patterns that represent points in multidimensional spaces of fixed dimensionality (Xiong and Yeung 2002). In these articles (Bahrepour et al. 2011; Singh and Borah 2012), researchers employ SOFM clustering algorithm for determining the intervals of the historical time series data sets by clustering them into different groups. This algorithm is developed by Kohonen (Kohonen 1990), which is a class of neural networks with neurons arranged in a low dimensional (often two-dimensional) structure, and trained by an iterative unsupervised or self-organizing procedure (Liao 2005). The SOFM converts the patterns of arbitrary dimensionality into response of one-dimensional or two-dimensional arrays of neurons, i.e., it converts a wide pattern space into a feature space. The neural network performing such a mapping is called feature map (Sivanandam and Deepa 2007).

(c) *For defuzzification operation*: (Singh and Borah (2013b) develop an ANN based architecture and hybridize this architecture with FTS model to defuzzify the fuzzified time series values. The neural network architecture as shown in Fig. 2.5 can be employed for this purpose. In this case, the arrangement of nodes in input layer can be done in the following sequence:

$$F(t-n), \ldots, F(t-2), F(t-1) \rightarrow F(t) \tag{2.4.1}$$

Here, each input node take the previous days $(t-n), \ldots, (t-2), (t-1)$ fuzzified time series values (e.g., $A_l, \ldots, A_m, A_n$) to predict one day (t) advance time series value "A_j". In Eq. 2.4.1, each "t" represent the day for considered fuzzified time series values.

2.4.2 RS: An Introduction

RS is a new mathematical tool proposed by Pawlak (Pawlak 1982). The RS concept (Cheng et al. 2010) is based on the assumption that with every associated object of the universe of discourse, some information objects characterized by the same information are indiscernible in the view of the available information about them. Any set of all indiscernible objects is called an elementary set and forms a basic granule of knowledge about the universe. Any union of elementary sets is referred to as a precise set; otherwise the set is rough. A fundamental advantage of RS theory is the ability to handle a category that cannot be sharply defined from a given knowledge base (Pattaraintakorn and Cercone 2008). Therefore, the RS theory is used

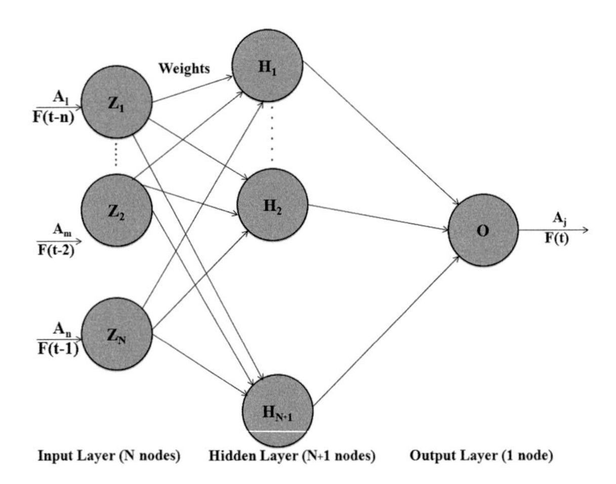

Fig. 2.5 ANN architecture for the nth-order FLRs

in attribute selection, rule discovery and various knowledge discovery applications as data mining, machine learning and medical diagnoses (Chen and Cheng 2013).

To understand the RS theory in-depth, we need to review some of the basic definitions as follows (Pawlak 1991):

U is a finite set of objects, i.e., $U = \{x_1, x_2, x_3, \ldots, x_n\}$. Here, each $x_1, x_2, x_3, \ldots, x_n$ represents the object.

Definition 2.4.1 (*Equivalence relation*). Let R be an equivalence relation over U, then the family of all equivalence classes of R is represented by U/R.

Definition 2.4.2 (*Lower approximation and upper approximation*). X is a subset of U, R is an equivalence relation, the lower approximation of X (i.e., $\underline{R}(X)$) and the upper approximation of X (i.e., $\overline{R}(X)$) is defined as follows:

$$\underline{R}(X) = \cup\{x \in U \mid [x]_R \subseteq X\} \tag{2.4.2}$$

$$\overline{R}(X) = \cup\{x \in U \mid [x]_R \cap X \neq \emptyset\} \tag{2.4.3}$$

The lower approximation comprises of all objects that completely belong to the set, and the upper approximation comprises all objects that possibly belong to the set.

Definition 2.4.3 (*Boundary region*). The set of all objects which can be decisively classified neither as members of X nor as members of non-X with respect to R is called the boundary region of a set X with respect to R, and denoted by RS_B.

$$RS_B = \overline{R}(X) - \underline{R}(X) \tag{2.4.4}$$

Based on the notions shown in Fig. 2.6, we can formulate the definitions of crisp set and RS as follows:

Definition 2.4.4 (*Crisp set*). A set X is called crisp (exact) with respect to R if and only if the boundary region of X is empty.

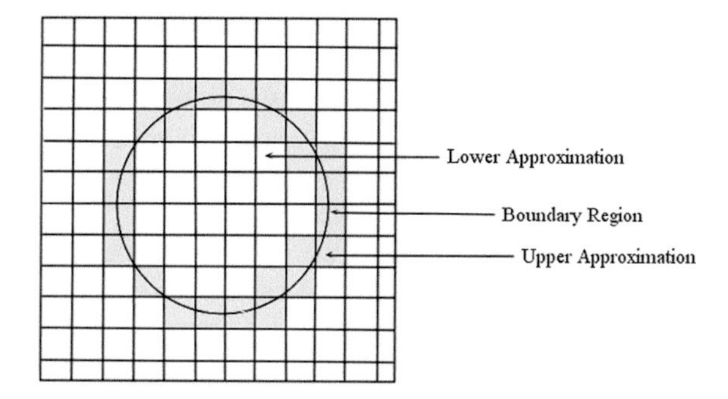

Fig. 2.6 Basic notations of the rough set

Definition 2.4.5 (*RS*). A set X is called rough (inexact) with respect to R if and only if the boundary region of X is nonempty.

The role of RS in FTS modeling approach is discussed below.

- *For rule induction*: In FTS model, each fuzzy set carries the information of occurrence of the historic event in the past. So, if these fuzzy sets would not be handled efficiently, then important information may be lost. Therefore, after generating the intervals, the historical time series data set is fuzzified, and can be used to prepare an information table. To mine reasonable rules from the information table, the RS based rule induction technique can be used, because the RS (Pawlak 1982) acts as a powerful tool for analyzing data and information tables. Teoh et al. 2008, 2009 employ this concept in FTS modeling approach to generate rules from the FLRs. The rules produced by RS rule induction method are in the form of "if-then" by combining a condition value (A_i) with several decision values ($A_j, A_k, \ldots, A_n$). For example, these decision values can be represented with "Then" as follows:

$$If\ (condition = A_i)\ Then\ (decision = A_j, A_k, \ldots, A_n) \tag{2.4.5}$$

2.4.3 EC: An Introduction

EC is a collection of problem solving techniques that includes paradigms such as Evolutionary Strategies, Evolutionary Programs and GAs (Bonissone 1997). GA concept was first proposed by (Holland (1975). All GAs contain three basic operators: reproduction, crossover, and mutation, where all three are analogous to their namesakes in genetics (Ross 2007). In GAs, a population consists of chromosomes and a chromosome consists of genes, where the number of chromosomes in a population is called the population size (Lee et al. 2007). In the following, we briefly review the basic concept of GA (Gen and Cheng 1997; Goldberg 1989; Sivanandam and Deepa 2007).

Step 1. *Create a random initial state*. An initial population is created from a random selection of solutions (chromosomes).

Step 2. *Evaluate fitness*. A value for fitness is assigned to each solution depending on how close it actually is to solving the problem.

Step 3. *Reproduce*. Those chromosomes with a higher fitness value are more likely to reproduce offspring.

Step 4. *Next generation*. If the new generation contains a solution that produces an output that is close enough or equal to the desired answer then the problem has been solved. Otherwise, iterate the whole process with the new generation.

PSO is a new algorithm of EC, which is applied to solve the bilevel programming problem (Wan et al. 2013). To deal with complicated optimization problem, recently many researchers hybridize this optimization technique with FTS modeling approach. In the following, we briefly review the basic concept of the PSO (Jiang et al. 2013; Lee 2006; Montalvo et al. 2008).

The PSO algorithm was first introduced by Eberhart and Kennedy 1995. It is a population-based evolutionary computation technique, which is inspired by the social behavior of animals such as bird flocking, fish schooling, and swarming theory (Eberhart and Shi 2001; Lin et al. 2010a, b). The PSO can be employed to solve many of the same kinds of problems as genetic algorithms (Kennedy and Eberhart 1995). The PSO algorithm is applied to a set of particles, where each particle has been assigned a randomized velocity. Each particle is then allowed to move towards the problem space. At each movement, each particle keeps track of its own best solution (fitness) and the best solution of its neighboring particles. The value of that fitness is called "*pbest*". Then each particle is attracted towards finding of the global best value by keeping track of overall best value of each particle, and its location (Trelea 2003). The particle which obtained the global fitness value is called "*gbest*".

At each step of optimization, velocity of each particle is dynamically adjusted according to its own experience and its neighboring particles, which is represented by the following equations:

$$Vel_{id,t} = \alpha \times Vel_{id,t} + M_1 \times R_{and} \times (PB_{id} - CP_{id,t}) + M_2 \times R_{and} \times (PG_{best} - CP_{id,t}) \tag{2.4.6}$$

The position of a new particle can be determined by the following equation:

$$CP_{id,t} = CP_{id,t} + Vel_{id,t} \tag{2.4.7}$$

where i represents the ith particle and d represents the dimension of the problem space. In Eq. 2.4.6, α represents the inertia weight factor; $CP_{id,t}$ represents the current position of the particle i in iteration t; PB_{id} denotes the previous best position of the particle i that experiences the best fitness value so far (*pbest*); PG_{best} represents the global best fitness value (*gbest*) among all the particles; R_{and} gives the random value in the range of [0, 1]; M_1 and M_2 represent the self-confidence coefficient and the

Algorithm 1 Standard PSO Algorithm

Step 1: Initialize all particles with random positions and velocities in the d-dimensional problem space.
Step 2: Evaluate the optimization fitness function of all particles.
Step 3: For each particle, compare its current fitness value with its $pbest$. If current value is better than $pbest$, then update $pbest$ value with the current value.
Step 4: For each particle, compare its fitness value with its overall previous best. If the current fitness value is better than $gbest$, then update $gbest$ value with the current best particle.
Step 5: For each particle, change the movement (velocity) and location (position) according to Eqs. 2.4.6 and 2.4.7.
Step 6: Repeat Step 2, until stopping criterion is met, usually a sufficiently $gbest$ value is obtained.

social coefficient, respectively; and $Vel_{id,t}$ represents the velocity of the particle i in iteration t. Here, $Vel_{id,t}$ is limited to the range $[-Vel_{max}, Vel_{max}]$, where Vel_{max} is a constant and defined by users. The steps for the standard PSO are presented in Algorithm 1.

The role of EC in FTS modeling approach is categorized below based on different functions.

(a) *For determination of optimal interval lengths using GA*: GA used in FTS modeling approach to arrive optimal interval lengths using certain genetic operators. In this case, some chromosomes are defined as the initial population based on the number of intervals, where each chromosome consists of genes. Initially each chromosome is randomly generated by the system. Then, the system randomly selects chromosomes and genes from the population to perform the crossover and mutation operations, respectively. The whole process is repeated until optimal interval lengths are achieved. The achievement of optimality can be measured with the performance measure parameters (refer to Sect. 2.6), such as AFER, MSE, etc. Based on this concept, researchers in these articles (Chen and Chung 2006a, b) presented the methods for forecasting the enrollments by hybridizing GA technique with FTS modeling approach. However, the basic difference between the models presented in these articles (Chen and Chung 2006a, b) is that the first model (Chen and Chung 2006b) is based on high-order FLRs, whereas the second model (Chen and Chung 2006a) is based on first-order FLRs. Similar to above approach, Lee et al. (2007, 2008) presented new methods for temperature and the TAIFEX forecasting based on two-factors high-orders FLRs.

(b) *For finding best intervals using PSO*: Recently, many researchers[8] show that appropriate selection of intervals also increases the forecasting accuracy of the model. Therefore, in order to get the optimal intervals, they used PSO algorithm in their proposed model.[9] They signify that PSO algorithm is more efficient and powerful than GA as applied by the researcher (Chen and Chung 2006b) in selection of proper intervals.

[8]References are: (Huang et al. 2011a, b; Kuo et al. 2009, 2010).

[9]References are: (Huang et al. 2011a, b; Kuo et al. 2009, 2010).

Algorithm 2 Type-2 FTS Forecasting Model

Step 1: Select Type-1 and Type-2 observations.
Step 2: Determine the universe of discourse of time series data set and partition it into different/equal lengths of intervals.
Step 3: Define linguistic terms for each of the interval.
Step 4: Fuzzify the time series data set of Type-1 and Type-2 observations.
Step 5: Establish the FLRs based on Definition 2.2.7.
Step 6: Construct the FLRGs based on Definition 2.2.8.
Step 7: Establish the relationships between FLRGs of both Type-1 and Type-2 observations, and map-out them to their corresponding day.
Step 8: Apply fuzzy operators (such as union or intersection) on mapped-out FLRGs of Type-1 and Type-2 observations, and obtain the fuzzified forecasting data.
Step 9: Defuzzify the forecasting data and compute the forecasted values.

(c) *For determination of membership values using PSO*: The PSO technique is first time employed by the researcher Aladag et al. (2012) to obtain the optimal membership values of the fuzzy sets in the fuzzy relationship matrix "R" (refer to Eq. 2.3.2). In this approach, first FCM clustering algorithm is used for fuzzification phase of time series data set.

2.5 Financial Forecasting and Type-2 FTS Models

The application of FTS in financial forecasting has attracted many researchers' attention in the recent years. Many any researchers focus on designing the models for the TAIEX[10] and the TIFEX[11] forecasting. Their applications are limited to deal with either one-factor or two-factors time series data sets. However, forecasting accuracy of financial data set can be improved by including more observations (e.g., *close, high*, and *low*) in the models. In Type-2 FTS modeling approach, observation that is handled by Type-1 FTS model can be termed as "main-factor / Type-1 observation", whereas observations that are handled by Type-2 FTS model can be termed as "secondary-factors / Type-2 observations". Later, both these observations are combined together to take the final decision. But, due to involvement of Type-2 observations with Type-1 observation, massive FLRGs are generated in Type-2 model. For this reason, Type-2 FTS model suffers from the burden of extra computation. Therefore, most of the researchers still use to prefer Type-1 FTS modeling approach for forecasting. But, as far as accuracy of forecasting is concerned, Type-2 FTS models produce better result than Type-1 FTS models. Basic steps involve in Type-2 FTS modeling approach that can deal with multiple observations together are presented in Algorithm 2.

[10]References are: (Cheng et al. 2013; Huarng and Yu 2012; Wei et al 2011; Yu and Huarng 2010).
[11]References are: (Aladag et al. 2012; Avazbeigi et al. 2010; Bai et al. 2011; Kuo et al. 2010).

Contributions of various researchers in Type-2 FTS models are presented below:

- *Huarng and Yu* (Huarng and Yu 2005) *model*: This model first time employs the Type-2 FTS concept in financial forecasting (TAIEX) by considering *close, high,* and *low* observations together. In this model, they suggested some improvement in Algorithm 2 as:
  (i) Introduction of union ($\vee$) and intersection ($\wedge$) operators. This operators are applied in **Step 8** of Algorithm 2. Both these operators are used to include Type-1 and Type-2 observations., and (ii) For defuzzification operation, they employ **Principal 1** and **Principal 2** (as discussed in Sect. 2.3) in **Step 9** of Algorithm 2.
- *Bajestani and Zare* (Bajestani and Zare 2011) *model*: This model is the enhancement of the model proposed by Huarng and Yu (2005). In this model, researchers employ the four changes as:
  (i) Using triangular fuzzy set with indeterminate legs and optimizing these triangular fuzzy sets. This improvement is applied in **Step 3** of Algorithm 2., (ii) Using indeterminate coefficient in calculating Type-2 forecasting. This improvement is applied in **Step 9** of Algorithm 2., (iii) Using center of gravity defuzzifier. This improvement is applied in **Step 9** of Algorithm 2., and (iv) Using 4-order Type-2 FTS. This improvement is applied in **Step 5** of Algorithm 2.
- *Lertworaprachaya et al.* (Lertworaprachaya et al. 2010) *model*: Based on these articles (Huarng and Yu 2005; Singh 2007b), a novel high-order Type-2 FTS model is proposed in this article (Lertworaprachaya et al. 2010). This model is divided into two parts: high-order Type-1 FTS forecasting and Type-2 FTS forecasting. The high-order Type-1 FTS model is employed to define the FLRs. This improvement is suggested in **Step 5** of Algorithm 2. The high-order FLRs can be defined based on Definition 2.2.9. Then the rules in the high-order Type-1 FTS is used in Type-2 FTS forecasting.
- *Singh and Borah* (Singh and Borah 2013a) *model*: This Type-2 FTS model can utilize multiple observations together in forecasting, which was the limitation of previous existing Type-2 FTS models. Detail discussion on this model is provided in Chap. 6.

2.6 Performance Measure Parameters

To assess the performance of the time series forecasting models (especially FTS models), researchers use numerous performance measure parameters, such as $AFER$, MSE, $RMSE$, $\bar{A}$, SD, U, TS, DA, δ_r, R, R^2, PP, etc. All these parameters and their statistical significance are presented in Table 2.1. In this table, each F_i and A_i is the forecasted and actual value of day/year i, respectively, and N is the total number of days/years to be forecasted.

Table 2.1 Performance measure parameters and its statistical significance

Parameter	Significant
$AFER = \frac{\lvert F_i - A_i \rvert / A_i}{N} \times 100\,\%$	Smaller value of $AFER$ indicates good forecasting
$MSE = \frac{\sum_{i=1}^{N}(F_i - A_i)^2}{N}$	Smaller value of MSE indicates good forecasting
$RMSE = \sqrt{\frac{\sum_{i=1}^{N}(F_i - A_i)^2}{N}}$	Smaller value of $RMSE$ indicates good forecasting
$\bar{A} = \frac{\sum_{i=1}^{N} A_i}{N}$	For a good forecasting, the observed mean should be close to the predicted mean
$SD = \sqrt{\frac{1}{N}\sum_{i=1}^{N}(A_i - \bar{A})^2}$	For a good forecasting, the observed SD should be close to the predicted SD
$U = \frac{A}{B}$	Here, $A = \sqrt{\sum_{i=1}^{N}(A_i - F_i)^2}$ and $B = \sqrt{\sum_{i=1}^{N} A_i^2} + \sqrt{\sum_{i=1}^{N} F_i^2}$ The U is bounded between 0 and 1, with values closer to 0 indicating good forecasting accuracy
$TS = \frac{R_{sfe}}{M_{ad}}$	A TS value between −4 and +4 indicates that the model is working correctly
	Here, $M_{ad} = \frac{\sum_{i=1}^{N} \lvert (F_i - A_i) \rvert}{N}$ and $R_{sfe} = \sum_{i=1}^{N}(F_i - A_i)$
	A $M_{ad} > 0$ indicates that forecasting model tends to under-forecast
	A $M_{ad} < 0$ indicates that forecasting model tends to over-forecast
$DA = \frac{1}{N-1}\sum_{i=1}^{N-1} a_i$	Here, $a_i = \begin{cases} 1, & (A_{i+1} - A_i)(F_{i+1} - A_i) > 0 \\ 0, & Otherwise \end{cases}$
	DA value is measured in % and its value closer to 100 indicates good forecasting
$\delta_r = \frac{\lvert F_i - A_i \rvert}{SD}$	A value of δ_r less than 1 indicates good forecasting
$R = \frac{n\sum A_i F_i - (\sum A_i)(\sum F_i)}{\sqrt{n(\sum A_i^2) - (\sum A_i)^2}\sqrt{n(\sum F_i^2) - (\sum F_i)^2}}$	A value of R greater than equal to 0.8 is generally considered as strong
	The R^2 lies between $0 < R^2 < 1$, and indicates the strength of the linear association between A_i and F_i
$PP = 1 - (RMSE/SD)$	A PP value greater than zero indicates good forecasting and vice-versa

2.7 Conclusion and Discussion

From 1994 onwards, numerous time series forecasting models have been proposed based on the FTS modeling approach.[12] Due to the uncertain nature of time series, scope of extensive applications in this domain raised simultaneously with the development of new algorithms and architectures. The FTS modeling approach is currently applied to a diverse range of fields from economy, population growth, weather forecasting, stock index price forecasting to pollution forecasting, etc. Various aspects of complexities arise in this research domain, if the number of factors in time series data sets is large. These complexities can be evolved in terms of (a) Determination of length of intervals, (b) Establishment of FLRs between different factors, and (c) Defuzzification of fuzzified time series values.

Present research in the FTS modeling approach mainly aims at designing algorithms for discretization of time series data set, rule generation from the fuzzified time series values, proposing techniques for defuzzification operation, and designing various hybridized based architectures for resolving complex decision making problems.

SC techniques comprise of ANN, RS, EC, and their hybridizations, have recently been employed to solve FTS modeling problems. They endeavor to provide us approximate results in a very cost effective manner, thereby reducing the time complexity. In this survey, a categorization has been presented based on utilization of different SC techniques with the FTS modeling approach along with basic architectures of different hybridized based FTS models.

Fuzzy sets are the oldest component of SC, which is known for representation of real time or uncertain events in a linguistic manner, and can take decisions very faster. ANNs are especially used in discovering the rules, and can establish a linear association between the inputs and outputs. RSs is mainly employed for extracting hidden patterns from the data in terms of rules. EC provides efficient search algorithms to select based intervals from the discretized time series data set, based on some evaluation criterion.

FTS-ANN hybridization exploits the features of both ANN and fuzzy sets in establishment of FLRs/linguistic rules, data discretization, and defuzzification of fuzzified time series data set. FTS-RS hybridization uses the features of both RS and fuzzy sets in discovering meaning full rules from the fuzzified time series data set, thereby employing these rules in defuzzification operation. FTS-EC hybridization utilizes the characteristics of both EC and fuzzy sets in the determination of optimal interval lengths of the discretized time series data set, which are further used to represent time series data set in terms of fuzzy sets/linguistic terms. From this survey, it is obvious that the research scope in FTS will be increased in the near future for its flexibility in representing real life problems in a very natural way. This study also describes elaborately different phases of the FTS modeling approach. Various research issues and challenges in the FTS modeling approach are presented in the subsequent section. All

[12]References are: (Egrioglu et al. 2010, 2011a; Sah and Degtiarev 2005; Wong et al. 2010; Yu 2005a).

these inclusions may help the researchers to identify: (a) What are the problems in the FTS modeling approach?, (b) How to resolve all these problems using heuristics approach?, and (c) How to employ different SC methodologies in the FTS modeling approach to improve its efficiency?

References

Aladag CH, Basaran MA, Egrioglu E, Yolcu U, Uslu VR (2009) Forecasting in high order fuzzy times series by using neural networks to define fuzzy relations. Expert Syst Appl 36(3):4228–4231
Aladag CH, Yolcu U, Egrioglu E (2010) A high order fuzzy time series forecasting model based on adaptive expectation and artificial neural networks. Math Comput Simul 81(4):875–882
Aladag CH, Yolcu U, Egrioglu E, Dalar AZ (2012) A new time invariant fuzzy time series forecasting method based on particle swarm optimization. Appl Soft Comput 12(10):3291–3299
Avazbeigi M, Doulabi SHH, Karimi B (2010) Choosing the appropriate order in fuzzy time series: a new N-factor fuzzy time series for prediction of the auto industry production. Expert Syst Appl 37(8):5630–5639
Bahrepour M, Akbarzadeh-T MR, Yaghoobi M, Naghibi-S MB (2011) An adaptive ordered fuzzy time series with application to FOREX. Expert Syst Appl 38(1):475–485
Bai E, Wong WK, Chu WC, Xia M, Pan F (2011) A heuristic time-invariant model for fuzzy time series forecasting. Expert Syst Appl 38(3):2701–2707
Bajestani NS, Zare A (2011) Forecasting TAIEX using improved type 2 fuzzy time series. Expert Syst Appl 38(5):5816–5821
Bang YK, Lee CH (2011) Fuzzy time series prediction using hierarchical clustering algorithms. Expert Syst Appl 38(4):4312–4325
Beale MH, Hagan MT, Demuth HB (2010) Neural network Toolbox 7. The MathWorks Inc, Natick, MA
Bonissone PP (1997) Soft computing: the convergence of emerging reasoning technologies. Soft Comput 1:6–18
Brockwell PJ, Davis RA (2008) Introduction to time series and forecasting, 2nd edn. Springer, New York
Castro LN, Timmis JI (2003) Artificial immune systems as a novel soft computing paradigm. Soft Comput 7:526–544
Chang JR, Lee YT, Liao SY, Cheng CH (2007) Cardinality-based fuzzy time series for forecasting enrollments. New Trends Appl Artif Intell, vol 4570. Springer, Berlin/Heidelberg, pp 735–744
Chatfield C (2000) Time-series forecasting. Chapman and Hall, CRC Press, Boca Raton
Chen SM (1996) Forecasting enrollments based on fuzzy time series. Fuzzy Sets Syst 81:311–319
Chen SM (2002) Forecasting enrollments based on high-order fuzzy time series. Cybern Syst: Int J 33(1):1–16
Chen SM, Chen CD (2011a) Handling forecasting problems based on high-order fuzzy logical relationships. Expert Syst Appl 38(4):3857–3864
Chen SM, Chen CD (2011b) Handling forecasting problems based on high-order fuzzy logical relationships. Expert Syst Appl 38(4):3857–3864
Chen SM, Chung NY (2006a) Forecasting enrollments of students by using fuzzy time series and genetic algorithms. Int J Inf Manag Sci 17(3):1–17
Chen SM, Chung NY (2006b) Forecasting enrollments using high-order fuzzy time series and genetic algorithms. Int J Intell Syst 21(5):485–501
Chen SM, Hwang JR (2000) Temperature prediction using fuzzy time series. IEEE Trans Syst Man Cybern Part B: Cybern 30:263–275
Chen SM, Tanuwijaya K (2011) Multivariate fuzzy forecasting based on fuzzy time series and automatic clustering techniques. Expert Syst Appl 38(8):10,594–10,605

Chen SM, Wang NY (2010) Fuzzy forecasting based on fuzzy-trend logical relationship groups. IEEE Trans Syst Man Cybern Part B: Cybern 40(5):1343–1358
Chen TL, Cheng CH, Teoh HJ (2008) High-order fuzzy time-series based on multi-period adaptation model for forecasting stock markets. Phys A: Stat Mech Appl 387(4):876–888
Chen TY (2012) A signed-distance-based approach to importance assessment and multi-criteria group decision analysis based on interval type-2 fuzzy set. Knowl Inf Syst
Chen YS, Cheng CH (2013) Application of rough set classifiers for determining hemodialysis adequacy in ESRD patients. Knowl Inf Syst 34:453–482
Cheng C, Chang J, Yeh C (2006) Entropy-based and trapezoid fuzzification-based fuzzy time series approaches for forecasting IT project cost. Technol Forecast Soc Change 73:524–542
Cheng CH, Cheng GW, Wang JW (2008a) Multi-attribute fuzzy time series method based on fuzzy clustering. Expert Syst Appl 34:1235–1242
Cheng CH, Wang JW, Li CH (2008b) Forecasting the number of outpatient visits using a new fuzzy time series based on weighted-transitional matrix. Expert Syst Appl 34(4):2568–2575
Cheng CH, Chen TL, Wei LY (2010) A hybrid model based on rough sets theory and genetic algorithms for stock price forecasting. Inf Sci 180(9):1610–1629
Cheng CH, Huang SF, Teoh HJ (2011) Predicting daily ozone concentration maxima using fuzzy time series based on a two-stage linguistic partition method. Comput Math Appl 62(4):2016–2028
Cheng CH, Wei LY, Liu JW, Chen TL (2013) OWA-based ANFIS model for TAIEX forecasting. Econ Modell 30:442–448
Czibula G, Czibula IG, Găceanu RD (2013) Intelligent data structures selection using neural networks. Knowl Inf Syst 34:171–192
Donaldson RG, Kamstra M (1996) Forecast combining with neural networks. J Forecast 15(1):49–61
Dote Y, Ovaska SJ (2001) Industrial applications of soft computing: a review. Proc IEEE 89(9):1243–1265
Eberhart R, Kennedy J (1995) A new optimizer using particle swarm theory. In: Proceedings of the sixth international symposium on micro machine and human science, Nagoya, pp 39–43
Eberhart R, Shi Y (2001) Particle swarm optimization: Developments, applications and resources. In: Proceedings of the IEEE international conference on evolutionary computation, Brisbane, Australia, pp 591–600
Egrioglu E, Aladag CH, Yolcu U, Basaran MA, Uslu VR (2009) A new hybrid approach based on SARIMA and partial high order bivariate fuzzy time series forecasting model. Expert Syst Appl 36(4):7424–7434
Egrioglu E, Aladag CH, Yolcu U, Uslu VR, Basaran MA (2010) Finding an optimal interval length in high order fuzzy time series. Expert Syst Appl 37(7):5052–5055
Egrioglu E, Aladag CH, Basaran MA, Yolcu U, Uslu VR (2011a) A new approach based on the optimization of the length of intervals in fuzzy time series. J Intell Fuzzy Syst 22(1):15–19
Egrioglu E, Aladag CH, Yolcu U, Uslu V, Erilli N (2011b) Fuzzy time series forecasting method based on Gustafson-Kessel fuzzy clustering. Expert Syst Appl 38(8):10355–10357
Estivill-Castro V (2002) Why so many clustering algorithms: a position paper. ACM SIGKDD Explor Newslett 4(1):65–75
Gangwar SS, Kumar S (2012) Partitions based computational method for high-order fuzzy time series forecasting. Expert Systems with Applications 39(15):12,158–12,164
Gaume E, Gosset R (2003) Over-parameterisation, a major obstacle to the use of artificial neural networks in hydrology? Hydrol Earth Syst Sci 7(5):693–706
Gen M, Cheng R (1997) Genetic algorithms and engineering design. Wiley, New York
Goldberg DE (1989) Genetic algorithm in search, optimization, and machine learning. Addison-Wesley, Massachusetts
Gondek D, Hofmann T (2007) Non-redundant data clustering. Knowl Inf Syst 12:1–24
Greenfield S, Chiclana F (2013) Accuracy and complexity evaluation of defuzzification strategies for the discretised interval type-2 fuzzy set. Int J Approximate Reasoning 54(8):1013–1033

Haykin S (1999) Neural Networks, a comprehensive foundation. Macmillan College Publishing Co., New York
Herrera-Viedma E, Cabrerizo FJ, Kacprzyk J, Pedrycz W (2014) A review of soft consensus models in a fuzzy environment. Information Fusion 17:4–13
Holland JH (1975) Adaptation in natural and artificial systems. MIT Press, Cambridge
Hornik K, Stinchcombe M, White H (1989) Multilayer feedforward networks are universal approximators. Neural Netw 2(5):359–366
Hsu YY, Tse SM, Wu B (2003) A new approach of bivariate fuzzy time series analysis to the forecasting of a stock index. Int J Uncertain Fuzziness Knowl-Based Syst 11(6):671–690
Huang YL, Horng SJ, He M, Fan P, Kao TW, Khan MK, Lai JL, Kuo IH (2011a) A hybrid forecasting model for enrollments based on aggregated fuzzy time series and particle swarm optimization. Expert Syst Appl 38(7):8014–8023
Huang YL, Horng SJ, Kao TW, Run RS, Lai JL, Chen RJ, Kuo IH, Khan MK (2011b) An improved forecasting model based on the weighted fuzzy relationship matrix combined with a PSO adaptation for enrollments. Int J Innov Comput Inf Control 7(7A):4027–4046
Huarng K (2001) Heuristic models of fuzzy time series for forecasting. Fuzzy Sets Systems 123:369–386
Huarng K, Yu HK (2005) A Type 2 fuzzy time series model for stock index forecasting. Phys A: Stat Mech Appl 353:445–462
Huarng K, Yu THK (2006a) The application of neural networks to forecast fuzzy time series. Phys A: Stat Mech Appl 363(2):481–491
Huarng K, Yu THK (2006b) Ratio-based lengths of intervals to improve fuzzy time series forecasting. IEEE Trans Syst Man Cybern Part B: Cybern 36(2):328–340
Huarng KH, Yu THK (2012) Modeling fuzzy time series with multiple observations. Int J Innov Comput Inf Control 8(10(B)):7415–7426
Huarng KH, Yu THK, Hsu YW (2007) A multivariate heuristic model for fuzzy time-series forecasting. IEEE Trans Syst Man Cybern Part B: Cybern 37:836–846
Hwang JR, Chen SM, Lee CH (1998) Handling forecasting problems using fuzzy time series. Fuzzy Sets Syst 100:217–228
Indro DC et al (1999) Predicting mutual fund performance using artificial neural networks. Omega 27(3):373–380
Jang JSR, Sun CT, Mizutani E (1997) Neuro-fuzzy and soft computing: a computational approach to learning and machine intelligence. Prentice-Hall, London
Jiang Y, Li X, Huang C, Wu X (2013) Application of particle swarm optimization based on CHKS smoothing function for solving nonlinear bilevel programming problem. Appl Math Comput 219(9):4332–4339
Jilani TA, Burney SMA (2008) A refined fuzzy time series model for stock market forecasting. Phys A 387(12):2857–2862
Kacprzyk J (2010) Advances in soft computing, vol 7095. Springer, Heidelberg
Kennedy J, Eberhart R (1995) Particle swarm optimization. In: IEEE international conference on neural networks, Perth, WA 4:1942–1948
Keogh E, Lin J (2005) Clustering of time-series subsequences is meaningless: implications for previous and future research. Knowl Inf Syst 8(2):154–177
Kohonen T (1990) The self organizing maps. In: Proceedings of the IEEE, vol 78, pp 1464–1480
Kuligowski RJ, Barros AP (1998) Experiments in short-term precipitation forecasting using artificial neural networks. Mon Weather Rev 126:470–482
Kumar K, Bhattacharya S (2006) Artificial neural network vs. linear discriminant analysis in credit ratings forecast: A comparative study of prediction performances. Review of Accounting and Finance 5(3):216–227
Kumar S (2004) Neural networks: a classroom approach. Tata McGraw-Hill Education Pvt. Ltd., New Delhi

Kuo IH, Horng SJ, Kao TW, Lin TL, Lee CL, Pan Y (2009) An improved method for forecasting enrollments based on fuzzy time series and particle swarm optimization. Expert Syst Appl 36(3, Part 2):6108–6117

Kuo IH, Horng SJ, Chen YH, Run RS, Kao TW, Chen RJ, Lai JL, Lin TL (2010) Forecasting TAIFEX based on fuzzy time series and particle swarm optimization. Expert Syst Appl 37(2):1494–1502

Law R (2000) Back-propagation learning in improving the accuracy of neural network-based tourism demand forecasting. Tour Manag 21(4):331–340

Lee HS, Chou MT (2004) Fuzzy forecasting based on fuzzy time series. Int J Comput Math 81(7):781–789

Lee LW, Wang LH, Chen SM, Leu YH (2006) Handling forecasting problems based on two-factors high-order fuzzy time series. IEEE Trans Fuzzy Syst 14:468–477

Lee LW, Wang LH, Chen SM (2007) Temperature prediction and TAIFEX forecasting based on fuzzy logical relationships and genetic algorithms. Expert Syst Appl 33(3):539–550

Lee LW, Wang LH, Chen SM (2008) Temperature prediction and TAIFEX forecasting based on high-order fuzzy logical relationships and genetic simulated annealing techniques. Expert Syst Appl 34(1):328–336

Lee ZY (2006) Method of bilaterally bounded to solution blasius equation using particle swarm optimization. Appl Math Comput 179(2):779–786

Lertworaprachaya Y, Yang Y, John R (2010) High-order Type-2 fuzzy time series. International conference of soft computing and pattern recognition. IEEE, Paris, pp 363–368

Li ST, Cheng YC (2007) Deterministic fuzzy time series model for forecasting enrollments. Comput Math Appl 53(12):1904–1920

Li ST, Cheng YC, Lin SY (2008) A FCM-based deterministic forecasting model for fuzzy time series. Comput Math Appl 56(12):3052–3063

Li ST, Kuo SC, Cheng YC, Chen CC (2011) A vector forecasting model for fuzzy time series. Appl Soft Comput 11(3):3125–3134

Li ST et al (2010) Deterministic vector long-term forecasting for fuzzy time series. Fuzzy Sets Syst 161(13):1852–1870

Liao TW (2005) Clustering of time series data - a survey. Pattern Recogn 38(11):1857–1874

Lin SY, Horng SJ, Kao TW, Huang DK, Fahn CS, Lai JL, Chen RJ, Kuo IH (2010a) An efficient bi-objective personnel assignment algorithm based on a hybrid particle swarm optimization model. Expert Syst Appl 37(12):7825–7830

Lin TL, Horng SJ, Kao TW, Chen YH, Run RS, Chen RJ, Lai JL, Kuo IH (2010b) An efficient job-shop scheduling algorithm based on particle swarm optimization. Expert Syst Appl 37(3):2629–2636

Liu HT (2007) An improved fuzzy time series forecasting method using trapezoidal fuzzy numbers. Fuzzy Optim Decis Making 6:63–80

Liu HT, Wei ML (2010) An improved fuzzy forecasting method for seasonal time series. Expert Syst Appl 37(9):6310–6318

Liu HT, Wei NC, Yang CG (2009) Improved time-variant fuzzy time series forecast. Fuzzy Optim Decis Making 8:45–65

Liu Y, Starzyk JA, Zhu Z (2008) Optimized approximation algorithm in neural networks without overfitting. IEEE Trans Neural Netw 19(6):983–995

Mencattini A, Salmeri M, Bertazzoni S, Lojacono R, Pasero E, Moniaci W (2005) Local meteorological forecasting by type-2 fuzzy systems time series prediction. IEEE International conference on computational intelligence for measurement systems and applications. Giardini Naxos, Italy, pp 20–22

Mitra S, Pal SK, Mitra P (2002) Data mining in soft computing framework: a survey. IEEE Trans Neural Netw 13(1):3–14

Montalvo I, Izquierdo J, Pérez R, Tung MM (2008) Particle swarm optimization applied to the design of water supply systems. Comput Math Appl 56(3):769–776

Ordonez C (2003) Clustering binary data streams with K-means. In: Proceedings of the 8th ACM SIGMOD workshop on Research issues in data mining and knowledge discovery. ACM Press, New York, USA, pp 12–19
Own CM, Yu PT (2005) Forecasting fuzzy time series on a heuristic high-order model. Cybern Syst: Int J 36(7):705–717
Pattaraintakorn P, Cercone N (2008) Integrating rough set theory and medical applications. Appl Math Lett 21(4):400–403
Pawlak Z (1982) Rough sets. Int J Comput Inf Sci 11(5):341–356
Pawlak Z (1991) Rough sets: theoretical aspects of reasoning about data. Kluwer Academic Publishers
Pedrycz W, Sosnowski Z (2005) C-Fuzzy decision trees. IEEE Trans Syst Man Cybern C Appl Rev 35(4):498–511
Piotrowski AP, Napiorkowski JJ (2012) A comparison of methods to avoid overfitting in neural networks training in the case of catchment runoff modelling. J Hydrol
Qiu W, Liu X, Li H (2011) A generalized method for forecasting based on fuzzy time series. Expert Syst Appl 38(8):10446–10453
Qiu W, Liu X, Wang L (2012) Forecasting shanghai composite index based on fuzzy time series and improved C-fuzzy decision trees. Expert Syst Appl 39(9):7680–7689
Ross TJ (2007) Fuzzy Logic with engineering applications. John Wiley and Sons, Singapore
Sah M, Degtiarev K (2005) Forecasting enrollment model based on first-order fuzzy time series. Prec World Acad Sci Eng Technol 1:132–135
Singh P, Borah B (2012) An effective neural network and fuzzy time series-based hybridized model to handle forecasting problems of two factors. Knowl Inf Syst 38(3):669–690
Singh P, Borah B (2013a) Forecasting stock index price based on M-factors fuzzy time series and particle swarm optimization. International Journal of Approximate Reasoning
Singh P, Borah B (2013b) High-order fuzzy-neuro expert system for daily temperature forecasting. Knowl-Based Syst 46:12–21
Singh SR (2007a) A robust method of forecasting based on fuzzy time series. Appl Math Comput 188(1):472–484
Singh SR (2007b) A simple method of forecasting based on fuzzy time series. Appl Math Comput 186(1):330–339
Singh SR (2007c) A simple time variant method for fuzzy time series forecasting. Cybernetics and Systems: An International Journal 38(3):305–321
Singh SR (2008) A computational method of forecasting based on fuzzy time series. Math Comput Simul 79(3):539–554
Singh SR (2009) A computational method of forecasting based on high-order fuzzy time series. Expert Syst Appl 36(7):10,551–10,559
Sivanandam SN, Deepa SN (2007) Principles of soft computing. Wiley India (P) Ltd., New Delhi
Song Q, Chissom BS (1993a) Forecasting enrollments with fuzzy time series - Part I. Fuzzy Sets Syst 54(1):1–9
Song Q, Chissom BS (1993b) Fuzzy time series and its models. Fuzzy Sets Syst 54(1):1–9
Song Q, Chissom BS (1994) Forecasting enrollments with fuzzy time series - Part II. Fuzzy Sets Syst 62(1):1–8
Szmidt E, Kacprzyk J, Bujnowski P (2014) How to measure the amount of knowledge conveyed by atanassov's intuitionistic fuzzy sets. Inf Sci 257:276–285
Tan PN, Steinbach M, Kumar V (2009) Introduction to data mining, 4th edn. Dorling Kindersley Publishing Inc, India
Taylor JW, Buizza R (2002) Neural network load forecasting with weather ensemble predictions. IEEE Trans Power Syst 17:626–632
Teoh HJ, Cheng CH, Chu HH, Chen JS (2008) Fuzzy time series model based on probabilistic approach and rough set rule induction for empirical research in stock markets. Data Knowl Eng 67(1):103–117

Teoh HJ, Chen TL, Cheng CH, Chu HH (2009) A hybrid multi-order fuzzy time series for forecasting stock markets. Expert Syst Appl 36(4):7888–7897
Trelea IC (2003) The particle swarm optimization algorithm: convergence analysis and parameter selection. Inf Process Lett 85(6):317–325
Tsai CC, Wu SJ (2000) Forecasting enrolments with high-order fuzzy time series. 19th International conference of the North American. Fuzzy Information Processing Society, Atlanta, GA, pp 196–200
Tsaur RC, Yang JCO, Wang HF (2005) Fuzzy relation analysis in fuzzy time series model. Comput Math Appl 49(4):539–548
Venkatesan C, Raskar SD, Tambe SS, Kulkarni BD, Keshavamurty RN (1997) Prediction of all India summer monsoon rainfall using error-back-propagation neural networks. Meteorol Atmos Phys 62:225–240
Wan Z, Wang G, Sun B (2013) A hybrid intelligent algorithm by combining particle swarm optimization with chaos searching technique for solving nonlinear bilevel programming problems. Swarm Evol Comput 8:26–32
Wei LY, Chen TL, Ho TH (2011) A hybrid model based on adaptive-network-based fuzzy inference system to forecast Taiwan stock market. Expert Syst Appl 38(11):13,625–13,631
Weigend A (1994) An overfitting and the effective number of hidden units. In: Mozer MC, Smolensky P, Weigend AS (eds) Proceedings of the 1993 Connectionist Models Summer School. Lawrence Erlbaum Associates, Hillsdale, NJ, pp 335–342
Wilson ID, Paris SD, Ware JA, Jenkins DH (2002) Residential property price time series forecasting with neural networks. Knowl-Based Syst 15(5–6):335–341
Wong WK, Bai E, Chu AWC (2010) Adaptive time-variant models for fuzzy-time-series forecasting. IEEE Trans Syst Man Cybern Part B: Cybern 40:453–482
Wu X et al (2008) Top 10 algorithms in data mining. Knowl Inf Syst 14:1–37
Xiong Y, Yeung DY (2002) Mixtures of ARMA models for model-based time series clustering. In: IEEE International conference on data mining, Los Alamitos, USA, pp 717–720
Yardimci A (2009) Soft computing in medicine. Appl Soft Comput 9(3):1029–1043
Yolcu U et al (2011) Time-series forecasting with a novel fuzzy time-series approach: an example for Istanbul stock market. Journal of Statistical Computation and Simulation
Yu HK (2005a) A refined fuzzy time-series model for forecasting. Phys A 346(3–4):657–681
Yu HK (2005b) Weighted fuzzy time series models for TAIEX forecasting. Phys A 349(3–4):609–624
Yu THK, Huarng KH (2010) A neural network-based fuzzy time series model to improve forecasting. Expert Syst Appl 37(4):3366–3372
Zadeh LA (1965) Fuzzy sets. Inf Control 8(3):338–353
Zadeh LA (1971) Similarity relations and fuzzy orderings. Inf Sci 3:177–200
Zadeh LA (1973) Outline of a new approach to the analysis of complex systems and decision processes. IEEE Trans Syst Man Cybern SMC-3:28–44
Zadeh LA (1975a) The concept of a linguistic variable and its application to approximate reasoning - I. Inf Sci 8(3):199–249
Zadeh LA (1975b) The concept of a linguistic variable and its application to approximate reasoning-III. Inf Sci 9(1):43–80
Zadeh LA (1989) Knowledge representation in fuzzy logic. IEEE Trans Knowl Data Eng 1(1):89–100
Zadeh LA (1994) Fuzzy logic, neural networks, and soft computing. Commun ACM 37(3):77–84
Zadeh LA (1997) The roles of fuzzy logic and soft computing in the conception, design and deployment of intelligent systems. In: Software agents and soft computing, pp 183–190
Zadeh LA (2002) From computing with numbers to computing with words - From manipulation of measurements to manipulation of perceptions. Int J Appl Math Comput Sci 12(3):307–324

Chapter 3
Efficient One-Factor Fuzzy Time Series Forecasting Model

The purpose of computing is insight, not numbers
By Richard W. Hamming (1915–1998)

Abstract In this chapter, we present a new model to handle four major issues of FTS forecasting, viz., determination of effective length of intervals, handling of FLRs, determination of weight for each FLR, and defuzzification of fuzzified time series values. To resolve the problem associated with the determination of length of intervals, this study suggests a new time series data discretization technique. After generating the intervals, the historical time series data set is fuzzified based on the FTS theory. Each fuzzified time series values are then used to create the FLRs. Most of the existing FTS models simply ignore the repeated FLRs without any proper justification. Since FLRs represent the patterns of historical events as well as reflect the possibility of appearances these type of patterns in the future. If we simply discard the repeated FLRs, then there may be a chance of information lost. Therefore, in this model, it is recommended to consider the repeated FLRs during forecasting. It is also suggested to assign weights on the FLRs based on their severity rather than their patterns of occurrences. For this purpose, a new technique is incorporated in the model. This technique determines the weight for each FLR based on the index of the fuzzy set associated with the current state of the FLR. To handle these weighted FLRs and to obtain the forecasted results, this study proposes a new defuzzification technique.

Keywords FTS · Enrollment · Temperature · Stock exchange · Discretization · Defuzzification

P. Singh, *Applications of Soft Computing in Time Series Forecasting*,
Studies in Fuzziness and Soft Computing 330, DOI 10.1007/978-3-319-26293-2_3

3.1 Background and Related Literature

To enhance the accuracy in forecasted values, many researchers recently proposed various one-factor FTS models. For example, Chen et el. (2007) proposed a new FTS model for stock price forecasting by employing the concept of fibonacci sequence. In these articles,[1] researchers proposed computational methods of forecasting based on the high-order FLRs to overcome the drawback of fuzzy first-order forecasting models (Chen 1996; Huarng 2001b). Singh (2007) proposed a new method in FTS forecasting. This model has the advantage that it minimizes the time of complicated computations of fuzzy relational matrices or to find the steady state of fuzzy relational matrix. Li and Cheng (2007) proposed a new fuzzy deterministic model to handle three major issues, *viz.*, to control the uncertainty in forecasting, to partition the intervals effectively, and to obtain the forecasted results with different interval lengths. Liu (2007) designed an improved FTS forecasting method in which the forecasted value is considered as a trapezoidal fuzzy number instead of a single-point value. That model produced better forecasting results than the conventional models.[2] Wong et al. (2010) proposed an adaptive time-variant model that automatically adapts the analysis window size of FTS based on the predictive accuracy in the training phase and uses heuristic rules to determine forecasting values in the testing phase. The experiment results presented show that the model achieves a significant improvement in forecasting accuracy over the existing models.[3]

3.2 Input Data Selection

The effectiveness of the proposed model is demonstrated by using three real-world data sets: (a) University enrollments data set of Alabama (Song and Chissom 1993b), (b) Daily average temperature data set of Taipei (Chen and Hwang 2000), and (c) Daily stock exchange price data set of SBI (Finance 2012). However, the university enrollments data set of Alabama is used for the model verification, whereas the daily average temperature data set of Taipei and the daily stock exchange price data set of SBI is used for the model validation.

---

[1]References are: (Chen 2002; Chen and Tanuwijaya 2011a; Chen et al. 2008; Gangwar and Kumar 2012; Singh 2009).

[2]References are: (Chen 1996; Hwang et al. 1998; Lee and Chou 2004).

[3]References are: (Chen 1996; Huarng 2001b; Singh 2008; Song and Chissom 1993a; Tsaur et al. 2005).

3.3 MBD Approach

In this section, we propose a new discretization approach referred to as "Mean Based Discretization(MBD)" for determining the universe of discourse of the historical time series data set and partitioning it into different lengths of intervals. To explain this approach, the university enrollments data set of Alabama (Song and Chissom 1993b), shown in Table 3.1, is employed. Each step of the approach is elucidated below.

Step 3.3.1 Compute the mean of a sample, $S = \{x_1, x_2, \ldots, x_n\}$ of n measurements as:

$$\bar{x} = \frac{\sum_{i=1}^{n} x_i}{n} \tag{3.3.1}$$

In this example, the sample S is the university enrollments data set of Alabama (Table 3.1). Now, mean of the sample S is obtained as:

$$\bar{x} = \frac{(13055 + 13563 + 13867 + \ldots + 19337 + 18876)}{22} = 16194$$

Step 3.3.2 Find the sub-sets of sample S such that:

$$A = \{x \in S | x \leq \bar{x}\} \tag{3.3.2}$$
$$B = \{x \in S | x \geq \bar{x}\} \tag{3.3.3}$$

From Table 3.1, find the elements of A and B, and arrange them in ascending order as:

Table 3.1 Enrollments data set of the University of Alabama

Year	Actual enrollment	Year	Actual enrollment
1971	13055	1982	15433
1972	13563	1983	15497
1973	13867	1984	15145
1974	14696	1985	15163
1975	15460	1986	15984
1976	15311	1987	16859
1977	15603	1988	18150
1978	15861	1989	18970
1979	16807	1990	19328
1980	16919	1991	19337
1981	16388	1992	18876

$$A = \{13055, 13563, 13867, 14696, \ldots, 15984\} \quad (3.3.4)$$
$$B = \{16388, 16807, 16859, 16919, \ldots, 19337\} \quad (3.3.5)$$

Step 3.3.3 Define boundaries for A and B as:

$$U_A = [A_{min}, \bar{x}] \quad (3.3.6)$$
$$U_B = [\bar{x}, B_{max}], \quad (3.3.7)$$

where U_A and U_B are the boundaries for sub-sets A and B respectively. Here, A_{min} and B_{max} represent the minimum and maximum values of sub-sets A and B respectively.

From 3.3.4, we can define the boundaries for A and B as:

$$U_A = [13055, 16194] \quad (3.3.8)$$
$$U_B = [16194, 19337] \quad (3.3.9)$$

Step 3.3.4 Compute factors for A and B as:

$$F_A = \sum_{i=2}^{n_A} (x_i - x_{i-1}), x_i \in A, n_A = |A| \quad (3.3.10)$$
$$F_B = \sum_{i=2}^{n_B} (x_i - x_{i-1}), x_i \in B, n_B = |B| \quad (3.3.11)$$

where F_A and F_B are the factors for A and B respectively.

From 3.3.4, we can obtain the factors for A and B as:

$$F_A = (13563 - 13055) + (13867 - 13563) + \ldots + (15984 - 15861) = 2929 \quad (3.3.12)$$
$$F_B = (16807 - 16388) + (16859 - 16807) + \ldots + (16919 - 16859) = 2949 \quad (3.3.13)$$

Step 3.3.5 Determine deciding factors for A and B as:

$$DF_A = F_A / n_A \quad (3.3.14)$$
$$DF_B = F_B / n_B \quad (3.3.15)$$

where DF_A and DF_B are the deciding factors for A and B respectively. Here, n_A and n_B represent the number of elements in A and B, respectively.

The deciding factors for A and B (from (3.3.12) and (3.3.12)) are:

$$DF_A = 2929/12 = 244.08 \simeq 245 \quad (3.3.16)$$

$$DF_B = 2949/8 = 368.63 \simeq 369 \quad (3.3.17)$$

Step 3.3.6 Partition the boundaries U_A and U_B into different lengths of intervals as:

$$u_i = [L(i), U(i)], i = 1, 2, 3, \ldots; 1 \leq U(i) < \bar{x}; u_i \in U_A; \quad (3.3.18)$$

where $L(i) = A_{min} + (i - 1) \times DF_A$, and $U(i) = A_{min} + i \times DF_A$.

$$v_i = [L(i), V(i)], i = 1, 2, 3, \ldots; 1 \leq V(i) < B_{max}; v_i \in U_B; \quad (3.3.19)$$

where $L(i) = \bar{x} + (i - 1) \times DF_B$, and $V(i) = \bar{x} + i \times DF_B$.

Based on Eq. 3.3.18, intervals for boundary U_A are:

$$u_1 = [13055, 13300], u_2 = [13300, 13545], \ldots, u_{12} = [15750, 15995]$$

Similarly, based on Eq. 3.3.19, intervals for boundary U_B are:

$$v_1 = [16194, 16563], v_2 = [16563, 16932], \ldots, v_9 = [19146, 19515]$$

Step 3.3.7 Allocate the data to their corresponding intervals.

Allocate the elements of A and B to their corresponding intervals obtained after partitioning the boundaries U_A and U_B, respectively. All these intervals along with their corresponding elements are shown in Table 3.2. Last column of Table 3.2 represents the mid-points of newly determined intervals. Intervals which do not cover historical data are discarded from the list.

3.4 The Proposed FTS Forecasting Model

The functionality of each phase of the model is explained below utilizing the university enrollments data set of Alabamal.

Phase 3.4.1 Partition the time series data set into different lengths of intervals.

Based on the MBD approach, the historical time series data set is partitioned into different lengths of intervals. The experimental results are presented in Table 3.2. Each interval of Table 3.2 is represented as: $b_1 = [13055, 13300]$, $b_2 = [13545, 13790], \ldots, b_{13} = [19146, 19515]$. For each interval, mid-point is determined, and stored for future consideration.

Table 3.2 New intervals, their corresponding elements and mid-points

Interval	Boundary for U_A	Corresponding element	Mid-point
b_1	[13055, 13300]	13055	13055
b_2	[13545, 13790]	13563	13563
b_3	[13790, 14035]	13867	13867
b_4	[14525, 14770]	14696	14696
b_5	[15015, 15260]	15145, 15163	15154
b_6	[15260, 15505]	15311, 15433, 15460, 15497	15425
b_7	[15505, 15750]	15603	15603
b_8	[15750, 15995]	15861, 15984	15923
Interval	Boundary for U_B	Corresponding element	Mid-point
b_9	[16194, 16563]	16388	16388
b_{10}	[16563, 16932]	16807,16859,16919	16862
b_{11}	[18039, 18408]	18150	18150
b_{12}	[18777, 19146]	18876,18970	18923
b_{13}	[19146, 19515]	19328,19337	19333

Phase 3.4.2 Define linguistic terms for each of the interval.

The historical time series data set is divided into 13 intervals (i.e., $b_1, b_2, \ldots,$ and b_{13}). Therefore, total 13 linguistic variables (i.e., $A_1, A_2, \ldots, A_{13}$) are defined. All these linguistic variables can be represented by fuzzy sets as shown below:

$$\begin{aligned}
A_1 &= 1/b_1 + 0.5/b_2 + 0/b_3 + \ldots + 0/b_{n-2} + 0/b_{n-1} + 0/b_{13},\\
A_2 &= 0.5/b_1 + 1/b_2 + 0.5/b_3 + \ldots + 0/b_{n-2} + 0/b_{n-1} + 0/b_{13},\\
A_3 &= 0/b_1 + 0.5/b_2 + 1/b_3 + \ldots + 0/b_{n-2} + 0/b_{n-1} + 0/b_{13},\\
A_4 &= 0/b_1 + 0/b_2 + 0.5/b_3 + \ldots + 0/b_{n-2} + 0/b_{n-1} + 0/b_{13},\\
&\vdots\\
A_{13} &= 0/b_1 + 0/b_2 + 0/b_3 + \ldots + 0/b_{n-2} + 0.5/b_{n-1} + 1/b_{13}.
\end{aligned}$$

Obtain the degree of membership of each year's enrollment value belonging to each A_i. Here, the maximum degree of membership of fuzzy set A_i occurs at interval b_i, and $1 \leq i \leq 13$.

Phase 3.4.3 Fuzzify the time series data set.

If one year's enrollment value belongs to the interval b_i, then the fuzzified enrollment value for that year is considered as A_i. For example, the enrollment value of year 1971 belongs to the interval b_1, hence it is fuzzified to A_1. In this way, the

Table 3.3 Fuzzified enrollments data set

Year	Actual enrollment	Fuzzified enrollment	Year	Actual enrollment	Fuzzified enrollment
1971	13055	A_1	1982	15433	A_6
1972	13563	A_2	1983	15497	A_6
1973	13867	A_3	1984	15145	A_5
1974	14696	A_4	1985	15163	A_5
1975	15460	A_6	1986	15984	A_8
1976	15311	A_6	1987	16859	A_{10}
1977	15603	A_7	1988	18150	A_{11}
1978	15861	A_8	1989	18970	A_{12}
1979	16807	A_{10}	1990	19328	A_{13}
1980	16919	A_{10}	1991	19337	A_{13}
1981	16388	A_9	1992	18876	A_{12}

Table 3.4 The first-order FLRs of the enrollments data set

FLR	FLR
$A_1 \rightarrow A_2$	$A_6 \rightarrow A_6$
$A_2 \rightarrow A_3$	$A_6 \rightarrow A_5$
$A_3 \rightarrow A_4$	$A_5 \rightarrow A_5$
$A_4 \rightarrow A_6$	$A_5 \rightarrow A_8$
$A_6 \rightarrow A_6$	$A_8 \rightarrow A_{10}$
$A_6 \rightarrow A_7$	$A_{10} \rightarrow A_{11}$
$A_7 \rightarrow A_8$	$A_{11} \rightarrow A_{12}$
$A_8 \rightarrow A_{10}$	$A_{12} \rightarrow A_{13}$
$A_{10} \rightarrow A_{10}$	$A_{13} \rightarrow A_{13}$
$A_{10} \rightarrow A_9$	$A_{13} \rightarrow A_{12}$
$A_9 \rightarrow A_6$	–

historical time series data set is fuzzified. The fuzzified enrollment values are shown in Table 3.3.

Phase 3.4.4 Establish the FLRs between the fuzzified time series values.

Based on Definition 2.2.7, we can establish FLRs between two consecutive fuzzified enrollment values. For example, in Table 3.3, fuzzified enrollment values for Years 1973 and 1974 are A_3 and A_4, respectively. So, we can establish a FLR between A_3 and A_4 as: $A_3 \rightarrow A_4$. In this way, we have obtained the first-order FLRs for the fuzzified enrollment values, which are presented in Table 3.4. Then, based on Definition 2.2.9, nth-order FLRs are established from the fuzzified enrollment values (shown in Table 3.3). In Table 3.5, the second-order FLRs are shown, where $n = 2$. Remaining high-order FLRs are obtained in a similar manner.

Phase 3.4.5 Create the FLRGs.

Table 3.5 The second-order FLRs of the enrollments data set

FLR	FLR
$A_1, A_2 \to A_3$	$A_9, A_6 \to A_6$
$A_2, A_3 \to A_4$	$A_6, A_6 \to A_5$
$A_3, A_4 \to A_6$	$A_6, A_5 \to A_5$
$A_4, A_6 \to A_6$	$A_5, A_5 \to A_8$
$A_6, A_6 \to A_7$	$A_5, A_8 \to A_{10}$
$A_6, A_7 \to A_8$	$A_8, A_{10} \to A_{11}$
$A_7, A_8 \to A_{10}$	$A_{10}, A_{11} \to A_{12}$
$A_8, A_{10} \to A_{10}$	$A_{11}, A_{12} \to A_{13}$
$A_{10}, A_{10} \to A_9$	$A_{12}, A_{13} \to A_{13}$
$A_{10}, A_9 \to A_6$	$A_{13}, A_{13} \to A_{12}$

Based on Definition 2.2.8, the FLRs can be grouped into a FLRG. For example, in Table 3.4, there are four FLRs with the same previous state, $A_6 \to A_6$, $A_6 \to A_7$, $A_6 \to A_6$, and $A_6 \to A_5$. Therefore, these FLRs can be grouped into the FLRG, $A_6 \to A_6, A_7, A_6, A_5$. Similarly, the FLRGs for the high-order FLRs can be created. The FLRGs for the second-order FLRs as shown in Table 3.5, are presented in Table 3.7.

In Table 3.6, repeated FLRs are also included in the FLRGs. For example, in Table 3.4, the FLR $A_6 \to A_6$ appears twice. Therefore, in Table 3.6, this repeated FLR also appears twice in the FLRG, $A_6 \to A_6, A_7, A_6, A_5$.

According to the proposed IBWT, indices in the current state of the FLRs represent the corresponding weights of the FLRs. For example, consider the following FLRs as:

$$A_6 \to A_6 \text{ with weight 6,}$$
$$A_6 \to A_7 \text{ with weight 7,}$$
$$A_6 \to A_6 \text{ with weight 6,}$$
$$A_6 \to A_5 \text{ with weight 5.}$$

Here, each FLR is assigned a weight 6, 7, 6, and 5, based on indices of the current state of the FLRs, which are A_6, A_7, and A_6, and A_5, respectively. Later, corresponding weight for each of the FLR is employed for defuzzification operation.

Phase 3.4.6 Calculate forecast and defuzzify the fuzzified time series data.

To defuzzify the fuzzified time series data set, we have proposed the "Index Based Defuzzification (IBDT)" technique. The advantage of using IBDT scheme is that it can deal with weighted FLRs efficiently and obtain the forecasted results. The technique consists of **Principle 1** and **Principle 2**. The steps involve in **Principle 1** are explained below.

Table 3.6 The second-order FLRGs of the enrollments data set

FLRGs	FLRGs
Group 1: $A_1, A_2 \rightarrow A_3$	Group 10: $A_{10}, A_{10} \rightarrow A_9$
Group 2: $A_2, A_3 \rightarrow A_4$	Group 11: $A_{10}, A_9 \rightarrow A_6$
Group 3: $A_3, A_4 \rightarrow A_6$	Group 12: $A_9, A_6 \rightarrow A_6$
Group 4: $A_4, A_6 \rightarrow A_6$	Group 13: $A_6, A_5 \rightarrow A_5$
Group 5: $A_5, A_5 \rightarrow A_8$	Group 14: $A_5, A_8 \rightarrow A_{10}$
Group 6: $A_6, A_6 \rightarrow A_7, A_5$	Group 15: $A_{10}, A_{11} \rightarrow A_{12}$
Group 7: $A_6, A_7 \rightarrow A_8$	Group 16: $A_{11}, A_{12} \rightarrow A_{13}$
Group 8: $A_7, A_8 \rightarrow A_{10}$	Group 17: $A_{12}, A_{13} \rightarrow A_{13}$
Group 9: $A_8, A_{10} \rightarrow A_{10}, A_{11}$	Group 18: $A_{13}, A_{13} \rightarrow A_{12}$

Table 3.7 The first-order FLRGs of the enrollments data set

FLRG	FLRG
Group 1: $A_1 \rightarrow A_2$	Group 8: $A_8 \rightarrow A_{10}, A_{10}$
Group 2: $A_2 \rightarrow A_3$	Group 9: $A_9 \rightarrow A_6$
Group 3: $A_3 \rightarrow A_4$	Group 10: $A_{10} \rightarrow A_{10}, A_9, A_{11}$
Group 4: $A_4 \rightarrow A_6$	Group 11: $A_{11} \rightarrow A_{12}$
Group 5: $A_5 \rightarrow A_5, A_8$	Group 12: $A_{12} \rightarrow A_{13}$
Group 6: $A_6 \rightarrow A_6, A_7, A_6, A_5$	Group 13: $A_{13} \rightarrow A_{13}, A_{12}$
Group 7: $A_7 \rightarrow A_8$	–

⋆ **Principle 1**: For forecasting year $Y(t)$, the fuzzified value for the year $Y(t-1)$ is required, where "t" is the current year which we want to forecast. The **Principle 1** is applicable only if there are more than one fuzzified values available in the current state. The steps under **Principle 1** are explained below.

Step 1 Obtain the fuzzified value for year $Y(t-1)$ as $A_i(i = 1, 2, 3 \ldots, n)$.

Step 2 Obtain the FLRG whose previous state is $A_i(i = 1, 2, 3 \ldots, n)$, and current state is $A_k, A_s, \ldots, A_n$, i.e., the FLRG is in the form of $A_i \rightarrow A_k, A_s, \ldots, A_n$.

Step 3 Find the intervals where the maximum membership values of the fuzzy sets $A_k, A_s, \ldots, A_n$ occur. Let these intervals be $a_k, a_s, \ldots, a_n$. All these intervals have the corresponding mid-points $C_k, C_s, \ldots, C_n$.

Step 4 Obtain the weights assigned to fuzzy sets $A_k, A_s, \ldots, A_n$. Let these weights be $W_k, W_s, \ldots, W_n$.

Step 5 Apply the following formula to calculate the forecasted value for year, $Y(t)$:

$$\begin{aligned} Forecast\,(t) = {} & C_k \times \left[\frac{W_k}{W_k + W_s + \ldots + W_n}\right] + \\ & C_s \times \left[\frac{W_s}{W_k + W_s + \ldots + W_n}\right] + \\ & \ldots \qquad + \\ & C_n \times \left[\frac{W_n}{W_k + W_s + \ldots + W_n}\right] \end{aligned} \tag{3.4.1}$$

★ **Principle 2**: This rule is applicable if there is only one fuzzified value in the current state. The steps involve in **Principle 2** are explained below.

Step 1 Obtain the fuzzified value for the year $Y(t-1)$ as $A_i (i = 0, 1, \ldots, n)$.

Step 2 Find the FLRG whose previous state is A_i, and current state is A_j, i.e., the FLRG is in the form of $A_i \rightarrow A_j$.

Step 3 Find the interval where the maximum value of the fuzzy set A_j occurs. Let this interval be a_j. This interval a_j have the corresponding mid-point C_j. For forecasting year, $Y(t)$, this mid-point C_j is the forecasted value.

Based on the proposed model, two examples are presented here to compute forecasted values of the university enrollments data set as follows:

[Example 1] Suppose, we want to forecast the enrollment on Year, 1985, using one-factor fuzzy logical time series. To compute this value, the fuzzified enrollment value of the previous state is required. For forecasting year, $Y(1985)$, the fuzzified enrollment value for year, $Y(1984)$ is obtained from Table 3.3, which is A_5. Then, obtain the FLRG whose previous state is A_5 from Table 3.7. In this case, the FLRG is $A_5 \rightarrow A_5, A_8$. Hence, **Principle 1** is applicable here, because the current state has two fuzzified values. Now, find the intervals where the maximum membership values of A_5 and A_8 occur from Table 3.2, which are b_5 and b_8, respectively. The corresponding mid-point and weight for the interval b_5 are 15154 and 5, respectively. The corresponding mid-point and weight for the interval b_8 are 15923 and 8, respectively. Now, based on Eq. 3.4.1, the forecasted enrollment for year, $Y(1985)$, can be computed as:

$$Forecast\,(1985) = 15154 \times \left[\frac{5}{5+8}\right] + 15923 \times \left[\frac{8}{5+8}\right] = 15627$$

[Example 2] Suppose, we want to forecast the enrollment on Year, 1973. To compute this value, the fuzzified enrollment value of the previous state is required. For forecasting year, $Y(1973)$, the fuzzified enrollment value for year, $Y(1972)$ is obtained from Table 3.3, which is A_2. Then, obtain the FLRG whose previous state is A_2 from Table 3.7. In this case, the FLRG is $A_2 \rightarrow A_3$. Hence, **Principle 2** is applicable here, because in the current state, only one fuzzified value is available. Now, find the interval where the maximum membership value for the fuzzy set A_3 occurs from Table 3.2, which is b_3. The interval b_3 has the mid-point 13867, which is the forecasted enrollment for year, $Y(1973)$.

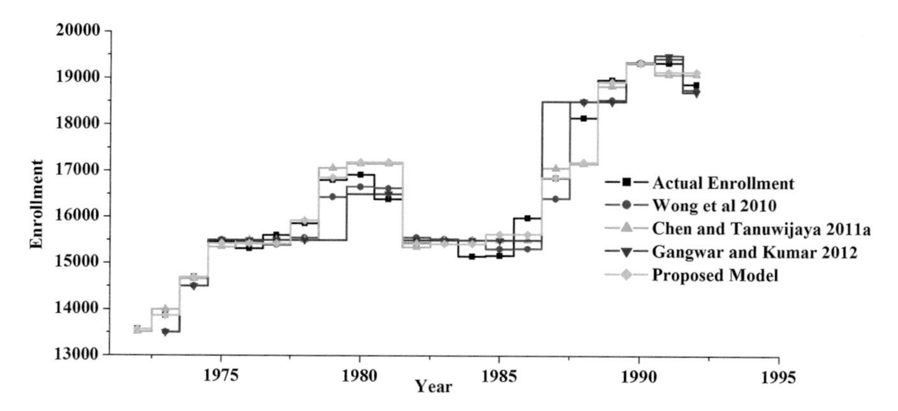

Fig. 3.1 Comparison curves of actual enrollment values and forecasted enrollment values

3.5 University Enrollments Forecasting

In this section, we present the empirical analysis of enrollments forecasting results obtained from the proposed model and the existing FTS models. To assess the performance of the proposed model, its forecasted results are compared with existing FTS models.[4] Table 3.8 presents a comparison of the forecasted results with the existing models in terms of $AFER$. From Table 3.8, it is observed that the computed value of $AFER$ is quite satisfactory for the proposed model as compared to the competing models, which depicts its superiority. A comparison graph is shown in Fig. 3.1, where the forecasted values obtained from the proposed model are compared with existing models (Chen and Tanuwijaya 2011a; Gangwar and Kumar 2012; Wong et al. 2010). In this figure, curves clearly indicate that the forecasted enrollments values obtained from the proposed model are very close to that of actual enrollments values.

To verify the superiority of the proposed model under different lengths of interval criteria, three existing forecasting models (Chen and Chen 2011; Huarng 2001a; Huarng and Yu 2006) are selected for comparison. A comparison of the forecasted results is presented in Table 3.9. We can see that the proposed model produces more precise results than the existing competing models under different lengths of intervals criteria.

Based on the number of intervals generated by the "MBD" approach and the high-order FLRs, performance of the model is verified next. Experimental results of the enrollments data set for different orders of the FLRs are presented in Table 3.10. To verify the superiority of the proposed model, it is compared with the existing high-order FTS models (Chen 2002; Chen and Tanuwijaya 2011a). Experimental results for these two models (Chen 2002; Chen and Tanuwijaya 2011a) are listed in the second and third columns of Table 3.10. From Table 3.10, it is obvious that

[4]References are: (Chen 1996; Chen and Tanuwijaya 2011a; Cheng et al. 2006, 2008; Gangwar and Kumar 2012; Lee and Chou 2004; Liu 2007; Qiu et al. 2011; Song and Chissom 1993a; Wong et al. 2010).

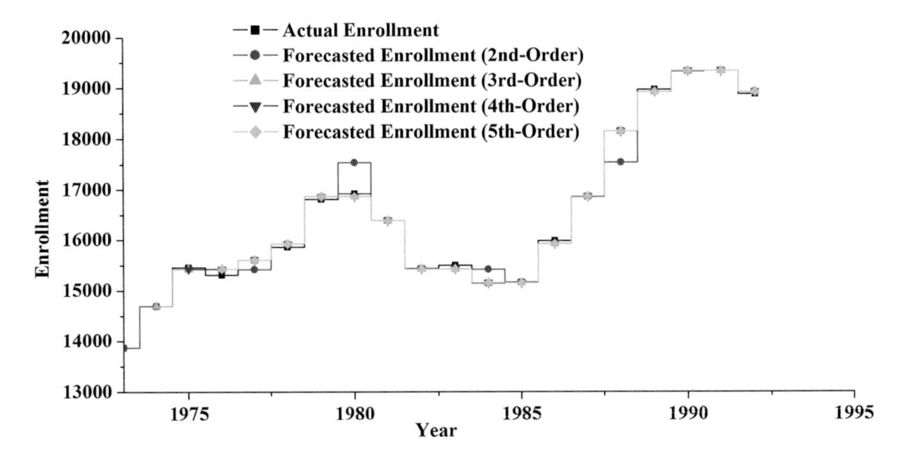

Fig. 3.2 Comparison curves of actual enrollment values and forecasted enrollment values based on the high-order FLRs

the forecasting accuracy of the proposed model is better than the ones presented in these articles (Chen 2002; Chen and Tanuwijaya 2011a). Graphical representation of the actual enrollment values and the forecasted enrollment values based on the high-order FLRs are shown in Fig. 3.2. Curves in Fig. 3.2 signify the effectiveness of the proposed model for forecasting enrollments based on the high-order FLRs.

To check the robustness of the proposed model, values of various statistical parameters as mentioned in Chap. 2 (see Sect. 2.6), are obtained. Experimental results are shown in Table 3.11. The values of parameters listed in Table 3.11 are based on the second-order FLRs. For comparison study, Gangwar and Kumar (Gangwar and Kumar 2012) model is selected, because its forecasting accuracy is better than the FTS models presented in these articles (Chen and Tanuwijaya 2011b; Liu 2007). In Table 3.11, the $RMSE$ value of the proposed model portrays very small error rate in comparison to Gangwar and Kumar (Gangwar and Kumar 2012) model. From the R value of the proposed model, it is obvious that the proposed model is more efficient than the considered model. The comparison of R^2 values between the proposed model, and the considered model indicates that the forecasted values are fitted very well with actual values in case of the proposed model. The smaller M_{ad} value of the proposed model indicates that the forecasted enrollment values obtained from the proposed model tend to be least under-forecast in comparison to the competing model. The PP values in Table 3.11 also signify the effectiveness of the proposed model in comparison to the Gangwar and Kumar (Gangwar and Kumar 2012) model. The TS values of both the models lie between the range ± 4, which indicate that both the models are working correctly. However, remaining statistical parameters indicate that the proposed model is more robust than the Gangwar and Kumar (Gangwar and Kumar 2012) model. Hence, from the above empirical analysis, it can be concluded that the proposed model is statistically significant in enrollments forecasting in comparison to the Gangwar and Kumar (Gangwar and Kumar 2012) model.

Table 3.8 A comparison of the existing models with proposed model

Year	Actual enrollment	Model Song and Chissom (1993a)	Model Chen (1996)	Model Lee and Chou (2004)	Model Cheng et al. (2006) (MEPA)	Model Cheng et al. (2006) (TFA)	Model Qiu et al. (2011)	Model Cheng et al. (2008)	Model Wong et al. (2010)	Model Chen and Tanuwijaya (2011a)	Model Liu (2007)	Model Gangwar and Kumar (2012)	Proposed model
1971	13055	–	–	–	–	–	–	–	–	–	–	–	–
1972	13563	14000	14000	14025	15430	14230	14195	14242	–	13512	13500	–	13563
1973	13867	14000	14000	14568	15430	14230	14424	14242	13500	13998	13800	13500	13867
1974	14696	14000	14000	14568	15430	14230	14593	14242	14500	14658	14700	14500	14696
1975	15460	15500	15500	15654	15430	15541	15589	15474.3	15500	15341	15600	15500	15425
1976	15311	16000	16000	15654	15430	15541	15645	15474.3	15466	15501	15400	15500	15420
1977	15603	16000	16000	15654	15430	15541	15634	15474.3	15392	15501	15750	15500	15420
1978	15861	16000	16000	15654	15430	16196	16100	15474.3	15549	15501	15400	15500	15923
1979	16807	16000	16000	16197	16889	16196	16188	16146.5	16433	17065	16800	–	16862
1980	16919	16813	16833	17283	16871	16196	17077	16988.3	16656	17159	17100	–	17192
1981	16388	16813	16833	17283	16871	17507	17105	16988.3	16624	17159	17100	16500	17192
1982	15433	16789	16833	16197	15447	16196	16369	16146.5	15556	15341	15300	15500	15425
1983	15497	16000	16000	15654	15430	15541	15643	15474.3	15524	15501	15750	15500	15420
1984	15145	16000	16000	15654	15430	15541	15648	15474.3	15497	15501	15400	15500	15420
1985	15163	16000	16000	15654	15430	15541	15622	15474.3	15305	15501	15300	15500	15627
1986	15984	16000	16000	15654	15430	15541	15623	15474.3	15308	15501	15750	–	15627
1987	16859	16000	16000	16197	16889	16196	16231	16146.5	16402	17065	16800	–	16862
1988	18150	16813	16833	17283	16871	17507	17090	16988.3	18500	17159	17100	18500	17192
1989	18970	19000	19000	18369	19333	18872	18325	19144	18534	18832	18900	18500	18923
1990	19328	19000	19000	19454	19333	18872	19000	19144	19345	19333	19200	19337	19333

(continued)

Table 3.8 (continued)

Year	Actual enrollment	Model Song and Chissom (1993a)	Model Chen (1996)	Model Lee and Chou (2004)	Model Cheng et al. (2006) (MEPA)	Model Cheng et al. (2006) (TFA)	Model Qiu et al. (2011)	Model Cheng et al. (2008)	Model Wong et al. (2010)	Model Chen and Tanuwi-jaya (2011a)	Model Liu (2007)	Model Gangwar and Kumar (2012)	Proposed model
1991	19337	19000	19000	19454	19333	18872	19000	19144	19423	19083	19050	19500	19136
1992	18876		19000		19333	18872	19000	19144	18752	19083	19050	18704	19136
AFER	–	3.22 %	3.11 %	2.67 %	2.75 %	2.66 %	2.66 %	2.40 %	1.52 %	1.54 %	1.33 %	1.27 %	1.19 %

Table 3.9 A comparison of the AFERs of the forecasting enrollments for different models

Year	Actual enrollment	Model Huarng (2001a) (Length of interval = 200)	Model Huarng (2001a) (Length of interval = 500)	Model Huarng and Yu (2006)	Model Chen and Chen (2011) (Length of interval = 1100)	MBD approach
1989	18970	18100	18250	18277	18810	18661
1990	19328	18900	18750	19128	19630	19327
1991	19337	19300	19250	19128	19450	18994
1992	18876	19300	19250	19128	19459	18994
$AFER$		2.31 %	2.30 %	1.78 %	1.52 %	1.07 %

Table 3.10 The AFERs to forecast the enrollments for different orders of the FLRs based on the existing models and the proposed model

Order	Model Chen (2002)	Model Chen and Tanuwijaya (2011a)	Proposed model
2	2.55 %	1.24 %	0.66 %
3	1.90 %	1.20 %	0.19 %
4	2.15 %	1.02 %	0.20 %
5	1.69 %	1.03 %	0.20 %

Table 3.11 Statistical analysis of the forecasted enrollments based on the existing model and the proposed model

Evaluation criterion	Model Gangwar and Kumar (2012)	Proposed model
$RMSE$	583.55	212.22
δ_r	0.14	0.07
R	0.95	0.99
R^2	0.89	0.98
PP	0.64	0.87
M_{ad}	393.15	110.55
TS	−0.5867	0.0629

3.6 Extended Applications

To further demonstrate the applicability of the proposed model, the daily average temperature data set in Taipei (Chen and Hwang 2000) from the period June (1996) to September (1996), and the daily stock exchange price of SBI (Finance 2012) from the period 6/1/2012 to 7/31/2012, are employed. The daily average temperature and stock exchange data sets are shown in Tables 3.12 and 3.13, respectively.

Table 3.12 A sample of daily average temperature data set from June (1996) to September (1996) in Taipei (Unit: °C)

Day	June	July	August	September
1	26.1	29.9	27.1	27.5
2	27.6	28.4	28.9	26.8
3	29.0	29.2	28.9	26.4
4	30.5	29.4	29.3	27.5
5	30.0	29.9	28.8	26.6
...	...	...	...	...
29	29.0	29.3	26.2	23.3
30	30.2	27.9	26.0	23.5
31	–	26.9	27.7	–

Table 3.13 The daily stock exchange price list of SBI from 6/5/2012 to 7/31/2012 (in Rupee)

Date (mm-dd-yy)	Actual price
6/5/2012	2080.25
6/6/2012	2159.45
6/7/2012	2167.85
6/8/2012	2180.05
6/11/2012	2164.55
6/12/2012	2206.90
6/13/2012	2222.25
6/14/2012	2154.25
...	...
7/26/2012	2017.15
7/27/2012	1941.20
7/31/2012	2005.20

The empirical results of forecasting the daily temperature and stock exchange price based on the first-order FLRs are presented in Tables 3.14 and 3.15, respectively. The forecasted results in terms of $RMSE$ indicate very small error rate for both the data sets. The computed values of δ_r (average) are less than 1 as shown in Tables 3.14 and 3.15. The R values between actual and forecasted values also indicate the efficiency of the proposed model. The R^2 values exhibit the strong linear association between actual and forecasted values. The PP values also indicate the efficiency of the proposed model. The M_{ad} values in Table 3.14 indicate that forecasted results of June, August and September tend to be slightly over-forecast. However, forecasted results of July (Table 3.14) and stock exchange price (Table 3.15) tend to be under-forecast. In Tables 3.14 and 3.15, it could be observed that TS values for both the data sets lie between the range ± 4, which indicate that the proposed model is working correctly.

Table 3.14 Empirical analysis of the daily average temperature forecasting from June (1996) to September (1996) in Taipei

Evaluation criterion	Forecasted temperature [June (I=19)]	Forecasted temperature [July (I=17)]	Forecasted temperature [August (I=20)]	Forecasted temperatur [September (I=18)]
$RMSE$	0.5708	0.6993	0.5628	0.5402
δ_r	0.3374	0.3533	0.5589	0.1816
R	0.8495	0.8421	0.8268	0.9658
R^2	0.7217	0.7092	0.6836	0.9328
PP	0.5163	0.4450	0.4315	0.7365
M_{ad}	0.3981	1.4367	0.3093	0.3723
TS	0.3162	0.2040	0.4164	0.2384

Table 3.15 Statistical analysis of the daily stock exchange price forecasting of SBI from 6/5/2012 to 7/31/2012 (Interval = 25)

Evaluation criterion	$RMSE$	δ_r	R	R^2	PP	M_{ad}	TS
Forecasted price	27.97	0.2673	0.9082	0.8249	0.57382	17.5449	0.3817

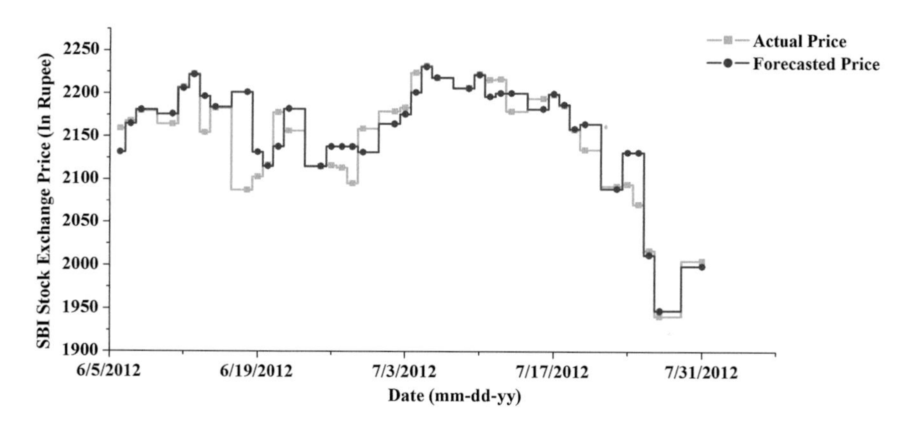

Fig. 3.3 Comparison curves of actual price and forecasted price for stock exchange data set of SBI

Hence, from the empirical analyzes, it is obvious that the proposed model is quite efficient in forecasting the daily temperature and stock exchange price with a very small error.

The curves of the actual temperature and the forecasted temperature, and the actual stock exchange price and the forecasted stock exchange price are shown in Figs. 3.3 and 3.4, respectively. It is obvious that the forecasted results are very close to that of actual values. Based on the number of intervals generated by the "MBD" approach

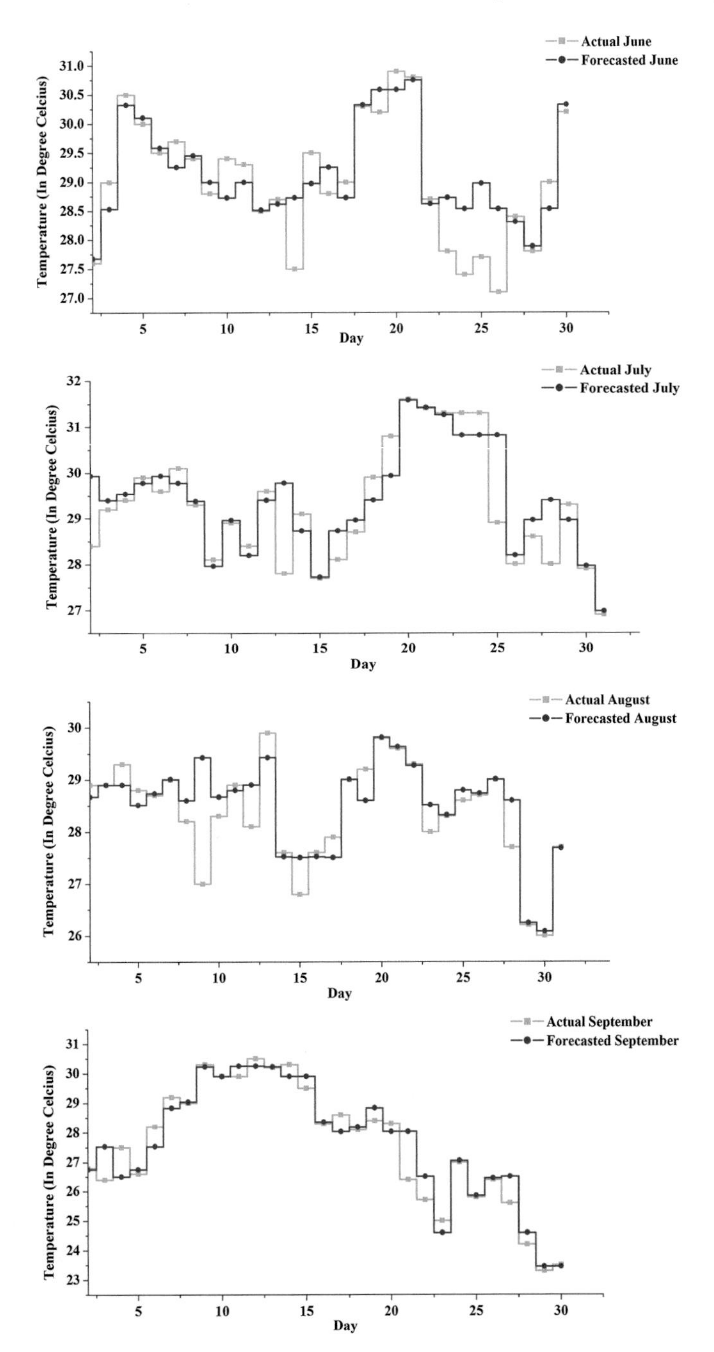

Fig. 3.4 Comparison curves of actual temperature and forecasted temperature for June, July, August and September (*top* to *bottom*)

Table 3.16 The AFERs to forecast the daily average temperature from June (1996) to September (1996) in Taipei for different orders of the FLRs based on the proposed model

Order	Forecasted temperature [June (I=19)] (%)	Forecasted temperature [July (I=17)] (%)	Forecasted temperature [August (I=20)] (%)	Forecasted temperature [September (I=18)] (%)
2	0.22	0.39	0.17	0.56
3	0.22	0.17	0.18	0.42
4	0.23	0.16	0.18	0.42
5	0.23	0.16	0.18	0.41
6	0.22	0.16	0.18	0.33
7	0.23	0.16	0.19	0.34
8	0.23	0.15	0.16	0.35
9	0.21	0.15	0.17	0.36

Table 3.17 The AFERs to forecast the daily stock exchange price from 6/5/2012 to 7/31/2012 for different orders of the FLRs based on the proposed model (Interval = 25)

Order	2	3	4	5	6	7	8	9
Forecasted price	0.59 %	0.60 %	0.61 %	0.62 %	0.57 %	0.57 %	0.59 %	0.55 %

and the high-order FLRs, performance of the model is evaluated next. Experimental results of forecasting the daily temperature and stock exchange price for different orders of the FLRs are presented in Tables 3.16 and 3.17, respectively. Empirical results in Tables 3.16 and 3.17 signify the effectiveness of the proposed model for forecasting the daily temperature and stock exchange price based on the high-order FLRs.

For comparison studies, various statistical models listed in article (Aladag et al. 2010) are simulated using PASW Statistics 18 (http://www.spss.com.hk/statistics/). A comparison of the forecasted accuracy among the existing statistical models and the proposed model for the daily average temperature data set and the daily stock exchange price data set are listed in Tables 3.18 and 3.19, respectively. The comparative analyzes clearly show that the proposed model outperforms over the considered statistical models in case of both the average temperature and stock exchange price data sets.

3.7 Discussion

In literature review, we have provided in-depth discussion of the recent works in FTS modeling approaches. Empirical analyzes indicate that the performance of these

Table 3.18 Empirical analysis of daily average temperature forecasting from June (1996) to September (1996) in Taipei based on the proposed model and the statistical models (in terms of the AFERs)

Model	Forecasted temperature (June) (%)	Forecasted temperature (July) (%)	Forecasted temperature (August) (%)	Forecasted temperature (September) (%)
Logarithmic regression	3.31	3.24	2.84	6.28
Inverse regression	3.11	3.23	2.87	6.35
Quadratic regression	3.19	3.32	2.81	2.58
Cubic regression	2.89	2.96	2.55	2.38
Compound regression	3.22	3.22	2.80	5.39
Power regression	3.32	3.23	2.84	6.29
S-curve regression	3.22	3.23	2.87	6.36
Growth regression	3.11	3.22	2.80	5.39
Exponential regression	3.22	3.22	2.80	5.39
Proposed model	1.40	1.54	1.11	1.38

Table 3.19 Empirical analysis of stock exchange price forecasting of SBI from 6/5/2012 to 7/31/2012 based on the proposed model and the statistical models (in terms of the AFERs)

Model	Forecasted stock exchange price (%)	Model	Forecasted stock exchange price (%)
Logarithmic regression	2.13	Power regression	2.14
Inverse regression	1.98	S-curve regression	1.97
Quadratic regression	2.04	Growth regression	2.40
Cubic regression	1.36	Exponential regression	2.40
Compound regression	2.40	Proposed model	0.82

existing models are below the satisfactory level. The main reasons of poor performance of these models are that they are designed on the following four assumptions: (a) determination of lengths of intervals in illogical way, (b) ignorance of repeated FLRs, (c) provide equal importance to each FLR, and (d) defuzzify the fuzzified time series values based on the Chen's centroid method (Chen 1996).

Therefore, in this chapter, we have presented a new FTS forecasting model that provides additional contributions or modifications in these assumptions, so that forecasting accuracy can be improved. The proposed model is designed on framework of Chen's model (Chen 1996) that can be represented by five major steps, as discussed in Sect. 2.3 (Chap. 2). Therefore, in the proposed model, we have required to modify these steps of the Chen's model (Chen 1996). These modifications are explained below.

Modification 1 In the proposed model, the "MBD" approach is incorporated in **Step 1** of the Chen's model (Chen 1996), which partitions the time series data set into different lengths of intervals.

Modification 2 In the proposed model, each FLR is assigned weight based on index of the fuzzy sets associated with the current state of FLR. The proposed scheme of weight assignment is referred to as "IBWT". The proposed model also considers the repeated FLRs during forecasting. These modifications contribute in **Step 4** of the Chen's model (Chen 1996).

Modification 3 To deal with the weighted FLRs and to compute the forecasted values, i.e., for the defuzzification operation, a new technique is introduced in **Step 5** of the Chen's model (Chen 1996), which is referred to as "IBDT".

For model verification, the University enrollments of Alabama from the period 1971–1992 are forecasted; whereas for model validation, the daily average temperature of Taipei and the stock exchange price of SBI are forecasted. In case of enrollments data set, forecasted results show that the proposed model outperforms than existing FTS models.[5]

The periodic forecasting of enrollments based on the proposed model from 1989–1992 also found to be very efficient from most of the other interval based approaches such as Average-based lengths intervals (Huarng 2001a), Distribution-based lengths intervals (Huarng 2001a), Ratio-based lengths of intervals (Huarng and Yu 2006), Chen and Chen Model (Length of interval= 1100) (Chen and Chen 2011). The empirical analyzes of the daily temperature and stock exchange price forecasting also signify the efficiency of the proposed model. The proposed model is verified and validated with the high-order FLRs. Superiority of the proposed model is also validated by comparing its forecasting accuracy with the various statistical models.

There is a limitation of the proposed model is that it can applicable only in one-factor time series data set. Hence, we have tried to make our model generalize enough so that it can deal with different kinds of one-factor time series data sets and can be used in various domains efficiently.

References

Aladag CH, Yolcu U, Egrioglu E (2010) A high order fuzzy time series forecasting model based on adaptive expectation and artificial neural networks. Math Comput Simul 81(4):875–882

Chen SM (1996) Forecasting enrollments based on fuzzy time series. Fuzzy Sets Syst 81:311–319

[5]References are: (Chen 1996; Chen and Tanuwijaya 2011a; Cheng et al. 2006, 2008; Gangwar and Kumar 2012; Lee and Chou 2004; Liu 2007; Qiu et al. 2011; Song and Chissom 1993a; Wong et al. 2010).

Chen SM (2002) Forecasting enrollments based on high-order fuzzy time series. Cybern Syst Int J 33(1):1–16
Chen SM, Chen CD (2011) Handling forecasting problems based on high-order fuzzy logical relationships. Expert Syst Appl 38(4):3857–3864
Chen SM, Hwang JR (2000) Temperature prediction using fuzzy time series. IEEE Trans Syst Man Cybern Part B Cybern 30:263–275
Chen SM, Tanuwijaya K (2011a) Fuzzy forecasting based on high-order fuzzy logical relationships and automatic clustering techniques. Expert Syst Appl 38(12):15425–15437
Chen SM, Tanuwijaya K (2011b) Multivariate fuzzy forecasting based on fuzzy time series and automatic clustering techniques. Expert Syst Appl 38(8):10594–10605
Chen TL, Cheng CH, Teoh HJ (2007) Fuzzy time-series based on fibonacci sequence for stock price forecasting. Phys: Stat Mech Appl 380:377–390
Chen TL, Cheng CH, Teoh HJ (2008) High-order fuzzy time-series based on multi-period adaptation model for forecasting stock markets. Phys A: Stat Mech Appl 387(4):876–888
Cheng C, Chang J, Yeh C (2006) Entropy-based and trapezoid fuzzification-based fuzzy time series approaches for forecasting IT project cost. Technol Forecast Social Change 73:524–542
Cheng CH, Cheng GW, Wang JW (2008) Multi-attribute fuzzy time series method based on fuzzy clustering. Expert Syst Appl 34:1235–1242
Finance Y (2012) Daily stock exchange price list of State Bank of India. http://in.finance.yahoo.com/
Gangwar SS, Kumar S (2012) Partitions based computational method for high-order fuzzy time series forecasting. Expert Syst Appl 39(15):12158–12164
Huarng K (2001a) Effective lengths of intervals to improve forecasting in fuzzy time series. Fuzzy Sets Syst 123:387–394
Huarng K (2001b) Heuristic models of fuzzy time series for forecasting. Fuzzy Sets Syst 123:369–386
Huarng K, Yu THK (2006) Ratio-based lengths of intervals to improve fuzzy time series forecasting. IEEE Trans Syst Man Cybern Part B Cybern 36(2):328–340
Hwang JR, Chen SM, Lee CH (1998) Handling forecasting problems using fuzzy time series. Fuzzy Sets Syst 100:217–228
Lee HS, Chou MT (2004) Fuzzy forecasting based on fuzzy time series. Int J Comput Math 81(7):781–789
Li ST, Cheng YC (2007) Deterministic fuzzy time series model for forecasting enrollments. Comput Math Appl 53(12):1904–1920
Liu HT (2007) An improved fuzzy time series forecasting method using trapezoidal fuzzy numbers. Fuzzy Optim Decis Mak 6:63–80
Qiu W, Liu X, Li H (2011) A generalized method for forecasting based on fuzzy time series. Expert Syst Appl 38(8):10446–10453
Singh SR (2007) A robust method of forecasting based on fuzzy time series. Appl Math Comput 188(1):472–484
Singh SR (2008) A computational method of forecasting based on fuzzy time series. Math Comput Simul 79(3):539–554
Singh SR (2009) A computational method of forecasting based on high-order fuzzy time series. Expert Syst Appl 36(7):10551–10559
Song Q, Chissom BS (1993a) Forecasting enrollments with fuzzy time series–Part I. Fuzzy Sets Syst 54(1):1–9
Song Q, Chissom BS (1993b) Fuzzy time series and its models. Fuzzy Sets Syst 54(1):1–9
Tsaur RC, Yang JCO, Wang HF (2005) Fuzzy relation analysis in fuzzy time series model. Comput Math Appl 49(4):539–548
Wong HL, Tu YH, Wang CC (2010) Application of fuzzy time series models for forecasting the amount of Taiwan export. Expert Syst Appl 37(2):1465–1470

Chapter 4
High-Order Fuzzy-Neuro Time Series Forecasting Model

Sometimes the questions are complicated and the answers are simple.

By Dr. Seuss (1904–1991)

Abstract In this chapter, we present a new model based on hybridization of FTS theory with ANN. In FTS models, lengths of intervals always affect the results of forecasting. So, for creating the effective lengths of intervals of the historical time series data set, a new "Re-Partitioning Discretization (RPD)" approach is introduced in the proposed model. Many researchers suggest that high-order fuzzy relationships improve the forecasting accuracy of the models. Therefore, in this study, we use the high-order fuzzy relationships in order to obtain more accurate forecasting results. Most of the FTS models use the current state's fuzzified values to obtain the forecasting results. The utilization of current state's fuzzified values (right hand side fuzzy relations) for prediction degrades the predictive skill of the FTS models, because predicted values lie within the sample. Therefore, for advance forecasting of time series, previous state's fuzzified values (left hand side of fuzzy relations) are employed in the proposed model. To defuzzify these fuzzified time series values, an ANN based architecture is developed, and incorporated in the proposed model.

Keywords FTS · High-order · Temperature · Stock exchange · Interval · FLR · ANN

4.1 Background and Related Literature

Many researchers have proposed various hybridization based models to solve complex problems in forecasting. For example, Hadavandi et al. (2010) presented a new approach based on genetic fuzzy systems and ANNs for building a stock price forecasting expert system to improve the forecasting accuracy. Cheng et al. (2010) proposed a new stock price forecasting model based on hybridization of GA with

P. Singh, *Applications of Soft Computing in Time Series Forecasting*,
Studies in Fuzziness and Soft Computing 330, DOI 10.1007/978-3-319-26293-2_4

RS theory. Kuo et al (2010) hybridized the particle swarm optimization with FTS to adjust the lengths of intervals in the universe of discourse. Aladag et al. (2009) introduced a new approach to define fuzzy relation in high order FTS using FFNN. Teoh et al. (2008) proposed a fuzzy-rough hybrid forecasting model, where rules (FLRs) are generated by RS algorithm. Pal and Mitra (2004) proposed a rough-fuzzy hybridization scheme for case generation. They used the fuzzy set theory for linguistic representation of patterns and then obtained the dependency rules by using the RS theory. For advance prediction of dwelling-fire occurrence in Derbyshire (United Kingdom), Yang et al. (2006) employed three approaches: logistic regression, ANN and GA. Keles et al. (2008) proposed a model for forecasting the domestic debt by Adaptive Neuro-Fuzzy Inference System. Chang et al. (2009) developed a hybrid model by integrating K-means clustering and fuzzy neural network to forecast the future sales of a printed circuit board factory. Huarng and Yu (2006), and Yu and Huarng (2008) presented a new hybrid model based on neural network and FTS to forecast TAIEX. Kuo et al. (2009) and Huang et al (2011) introduced a new enrollments forecasting model based on hybridization of FTS and PSO.

4.2 Architecture of the ANN

In this study, we use BPNN algorithm (Roy and Chakraborty 2013) to compute forecasted defuzzified value of the fuzzified time series data. Paradigms adopted for building the basic architecture for the proposed neural network is explained below.

Designing the right neural network architecture is a heuristic based approach and also a very time consuming process. The performance of the neural network architecture depends on number of layers, number of nodes in each layer and number of interconnection links with the nodes Wilson et al. (2002). Since, a neural network with more than three layers generate arbitrarily complex decision regions, a single hidden layer with one input layer and one output layer is considered here in designing the architecture. The number of nodes in input layer will depend on order of FLRs. For example, for third-order FLR, there would be three nodes in input layer; for fourth-order FLR, there would be four nodes in input layer, and so on. The minimum number of nodes in hidden layer is determined by the following equation:

$$Hidden_{nodes} = Input_{nodes} - 1, \tag{4.2.1}$$

where $Hidden_{nodes}$ and $Input_{nodes}$ represent the number of nodes in hidden and input layers respectively. A neural network architecture for the fifth-order FLRs is shown in Fig. 4.1.

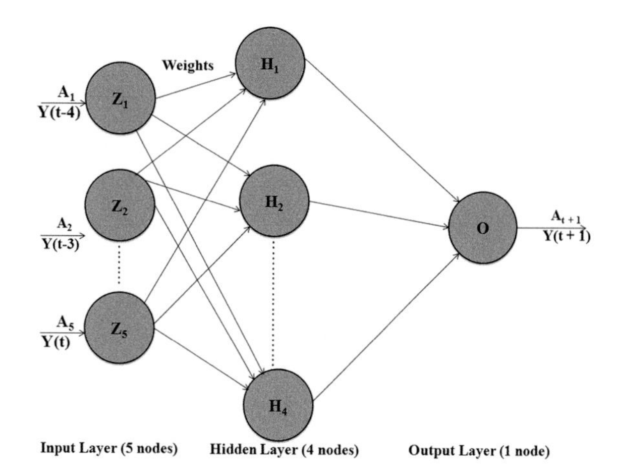

Fig. 4.1 A BPNN architecture for the fifth-order FLRs

The neural network as shown in Fig. 4.1 have 5 nodes (I_i, $i = 1, 2, \ldots, 5$) in input layer. The arrangement of nodes in input layer is done in the following sequence:

$$Y(t-4), Y(t-3), Y(t-2), Y(t-1), Y(t) \rightarrow Y(t+1) \tag{4.2.2}$$

Here, each input node takes the previous day's ($t-4, t-3, \ldots, t$) fuzzified time series values ($e.g.$, $A_1, A_2, \ldots, A_5$) to predict one day ($t+1$) advance daily temperature value "A_{t+1}". In Eq. 4.2.2, each "t" represents the day for considered fuzzified time series values.

4.3 Input Data Selection

For verifying the model, the daily average temperature data set from June 1996 to September 1996 in Taipei is employed. A sample of the data set is shown in Table 4.1. The complete set of data can be obtained from the Chen and Hwang's article (Chen and Hwang 2000). Taipei is situated at the northern tip of the island of China. It is the political, economic, and cultural center of China. So, advance prediction of daily temperature of Taipei is very advantageous for the inhabitants of Taipei.

4.4 RPD Approach

In this section, we propose a new discretization approach referred to as "Re-Partitioning Discretization (RPD)" for determining the universe of discourse of the historical time series data set and partitioning it into different lengths of

Table 4.1 A Sample of historical data of the daily average temperature from June 1996 to September 1996 in Taipei (Chen and Hwang 2000) (Unit: °C)

Day	Month			
	June	July	August	September
1	26.1	29.9	27.1	27.5
2	27.6	28.4	28.9	26.8
3	29.0	29.2	28.9	26.4
...	...	...	...	...
26	27.1	28.0	28.7	26.4
27	28.4	28.6	29.0	25.6
28	27.8	28.0	27.7	24.2
29	29.0	29.3	26.2	23.3
30	30.2	27.9	26.0	23.5
31	–	26.9	27.7	–

intervals.To explain this approach, the daily average temperature data set from June 1, 1996 to June 30, 1996, shown in Table 4.1, is employed. Each step of the approach is explained below.

Step 4.4.1 Compute range (R) of a sample, $S = \{x_1, x_2, \ldots, x_n\}$ as:

$$R = Max_{value} - Min_{value}, \tag{4.4.1}$$

where Max_{value} and Min_{value} are the maximum and minimum values of S respectively.

From Table 4.1, Max_{value} and Min_{value} for the June temperature data set (S) are 30.9 and 26.1 respectively. Therefore, the range R for this data set is computed as:

$$R = 30.9 - 26.1 = 4.8$$

Step 4.4.2 Split the data range R into M equally spaced classes, where M can be defined as Sturges (1926):

$$M = [1 + log_2^n], \tag{4.4.2}$$

where n is the size of the sample S.

Based on Eq. 4.4.2, we can compute M as:

$$M = 1 + \frac{1.477}{0.3010} = 5.907, \text{ where sample size n } = 30$$

Step 4.4.3 Obtain width of an interval (H) as:

$$H = \frac{R}{M} \tag{4.4.3}$$

Based on Eq. 4.4.3, we can calculate the width as:

$$H = \frac{4.8}{5.907} = 0.8126$$

Step 4.4.4 Define the universe of discourse U of the sample S as:

$$U = [L_b, U_b], \tag{4.4.4}$$

where $L_b = Min_{value} - H$, and $U_b = Max_{value} + H$.

Based on Table 4.1, we have the universe of discourse of the sample S as:

$$U = [26.1 - 0.8126, 30.9 + 0.8126] = [25.287, 31.713]$$

Step 4.4.5 Compute mid-point (U_{mid}) of the universe of discourse U as:

$$U_{mid} = \frac{L_b + U_b}{2} \tag{4.4.5}$$

The U_{mid} of the sample S is obtained as:

$$U_{mid} = \frac{25.287 + 31.713}{2} = 28.5$$

Step 4.4.6 Find sub-sets of the sample S such that:

$$A = \{x \in S | x \leq U_{mid}\} \tag{4.4.6}$$

$$B = \{x \in S | x \geq U_{mid}\} \tag{4.4.7}$$

From Table 4.1, we have obtained the elements of A and B as:

$$A = \{26.1, 27.1, 27.4, 27.5, 27.6, 27.7, 27.8, 27.8, 28.4, 28.5\}$$

$$B = \{28.7, 28.7, 28.8, 28.8, 29, 29, 29, 29.3, 29.4, 29.4, 29.5, 29.5, 29.7, 30, 30.2, 30.2, 30.3, 30.5, 30.8, 30.9\}$$

Step 4.4.7 Define sub-boundaries for A and B as:

$$U_A = [A_{min}, A_{max}] \tag{4.4.8}$$
$$U_B = [B_{min}, B_{max}], \tag{4.4.9}$$

where U_A and U_B are the sub-boundaries for A and B respectively. Here, A_{min} and A_{max} represent the minimum and maximum values of the sub-set A respectively. Similarly, B_{min} and B_{max} represent the minimum and maximum values of the sub-set B respectively.

From Definition 4.4.7, we can define the sub-boundaries for A and B as:

$$U_A = [26.1, 28.5] \tag{4.4.10}$$
$$U_B = [28.7, 30.9] \tag{4.4.11}$$

Step 4.4.8 Determine deciding factors for A and B as:

$$DF_A = \frac{A_{max} - A_{min}}{N_A} \tag{4.4.12}$$

$$DF_B = \frac{B_{max} - B_{min}}{N_B} \tag{4.4.13}$$

where DF_A and DF_B are the deciding factors for A and B respectively. Here, N_A and N_B represent the total number of elements of A and B respectively.

From Eqs. 4.4.12 and 4.4.13, the deciding factors for A and B are:

$$DF_A = \frac{28.5 - 26.1}{10} = 0.24 \tag{4.4.14}$$
$$DF_B = \frac{30.9 - 28.7}{20} = 0.11 \tag{4.4.15}$$

Step 4.4.9 Partition the sub-boundaries U_A and U_B into different lengths of intervals as:

$$u_i = [L(i), U(i)], i = 1, 2, 3, \ldots; 1 \leq U(i) < A_{max}; u_i \in U_A; \tag{4.4.16}$$

where $L(i) = A_{min} + (i - 1) \times DF_A$, and $U(i) = A_{min} + i \times DF_A$.

$$v_i = [M(i), V(i)], i = 1, 2, 3, \ldots; 1 \leq V(i) < B_{max}; v_i \in U_B; \tag{4.4.17}$$

where $M(i) = B_{min} + (i - 1) \times DF_B$, and $V(i) = B_{min} + i \times DF_B$.

Table 4.2 A sample of intervals and their corresponding elements for the June daily temperature data set

Interval for U_A	Corresponding element	Mid-point
[26.1, 26.34]	(26.1)	26.22
[27.06, 27.30]	(27.1)	27.18
[27.30, 27.54]	(27.4, 27.5)	27.42
...	...	...
[28.26, 28.50]	(28.4, 28.5)	28.38
Interval for U_B	Corresponding element	Mid-point
[28.70, 28.81]	(28.7, 28.7, 28.8, 28.8)	28.755
[28.92, 29.03]	(29, 29, 29)	28.975
[29.25, 29.36]	(29.3)	29.305
...	...	...
[30.46, 30.57]	(30.5)	30.515
[30.79, 30.90]	(30.8, 30.9)	30.845

Based on Eqs. 4.4.16 and 4.4.17, intervals for the sub-boundary U_A and U_B are:

$$u_1 = [26.1, 26.34], u_2 = [27.06, 27.30], \ldots, u_6 = [28.26, 28.50]$$
$$v_1 = [28.70, 28.81], v_2 = [28.92, 29.03], \ldots, v_{11} = [30.79, 30.90]$$

Step 4.4.10 Allocate the elements to their corresponding intervals.

Assign the elements of A and B to their corresponding intervals obtained after partitioning the boundaries U_A and U_B respectively. All these intervals along with their corresponding elements are shown in Table 4.2. Last column of Table 4.2 represents mid-points of the intervals. Intervals which do not cover historical data are discarded from the list. Intervals for the remaining three months July, August and September as shown in Table 4.1 are obtained in a similar way.

4.5 High-Order Fuzzy-Neuro Time Series Forecasting Model

Most of the existing FTS models use the five common steps, as discussed in Sect. 2.3 (Chap. 2), to deal with the forecasting problems. In this chapter, an improved FTS forecasting model is proposed, which is based on the hybridization of FTS theory with ANN. This model also employs the high-order FLRs to obtain the forecasting results. Therefore, these five steps are modified, which can be represented by the following steps:

1. Select the time series data set. Then, determine the universe of discourse of time series data set and partition it into different lengths of intervals based on "RPD" approach.
2. Define linguistic terms for each of the interval.
3. Fuzzify the time series data set.
4. Establish the high-order FLRs.
5. Calculate defuzzified forecasted value of the fuzzified time series data based on the BPNN architecture.

We apply the proposed model to forecast the daily temperature of Taipei from June, 1996 to September, 1996. This model is trained with the June daily temperature data set. Each phase of the training process is explained below.

Phase 4.5.1 Divide the universe of discourse into different lengths of intervals.

Define the universe of discourse U for the June temperature data set. Based on Eq. 4.4.4, we have defined the universe of discourse U as $U = [25.287, 31.713]$. Then, based on the RPD approach, the universe of discourse U is partitioned into n different lengths of intervals: $a_1, a_2, a_3, \ldots, a_n$. The results are presented in Table 4.2.

Phase 4.5.2 Define linguistic terms for each of the interval. Assume that the historical time series data set is distributed among n intervals (i.e., $a_1, a_2, \ldots,$ and a_n). Therefore, define n linguistic variables $A_1, A_2, \ldots, A_n$, which can be represented by fuzzy sets, as shown below:

$$\begin{aligned}
A_1 &= 1/a_1 + 0.5/a_2 + 0/a_3 + \ldots + 0/a_{n-2} + 0/a_{n-1} + 0/a_n,\\
A_2 &= 0.5/a_1 + 1/a_2 + 0.5/a_3 + \ldots + 0/a_{n-2} + 0/a_{n-1} + 0/a_n,\\
A_3 &= 0/a_1 + 0.5/a_2 + 1/a_3 + \ldots + 0/a_{n-2} + 0/a_{n-1} + 0/a_n,\\
&\vdots\\
A_j &= 0/a_1 + 0/a_2 + 0/a_3 + \ldots + 0/a_{n-2} + 0.5/a_{n-1} + 1/a_n. \qquad (4.5.1)
\end{aligned}$$

Obtain the degree of membership of each day's temperature value belonging to each A_i. Here, maximum degree of membership of fuzzy set A_i occurs at interval a_i, and $1 \leq i \leq n$.

Phase 4.5.3 Fuzzify the historical time series data. If one day's datum belongs to the interval a_i, then it is fuzzified into A_i, where $1 \leq i \leq n$.

Table 4.3 A sample of fuzzified historical data set for the June daily temperature

Day	June	Fuzzified temperature	Mid-point
1	26.1	A_1	26.22
2	27.6	A_4	27.66
3	29	A_8	28.975
4	30.5	A_{16}	30.515
5	30	A_{13}	29.965
6	29.5	A_{11}	29.525
...	...	...	...
28	27.8	A_5	27.9
29	29	A_8	28.975
30	30.2	A_{14}	30.185

[Explanation] If one day's temperature value belongs to the interval a_i, then the fuzzified temperature value for that day is considered as A_i. For example, the temperature value of June 1, 1996 belongs to the interval a_1, so it is fuzzified to A_1. In this way, we have fuzzified historical time series data set. A sample of fuzzified temperature values are shown in Table 4.3.

Phase 4.5.4 Establish the high-order FLRs between the fuzzified daily temperature values.

[Explanation] Based on Definition 2.2.9, we have established the fifth-order FLRs between the fuzzified daily temperature values. For example, in Table 4.3, the fuzzified daily temperature values for days $1, 2, 3, 4, 5$ and 6 are A_1, A_4, A_8, A_{16}, A_{13} and A_{11}, respectively. Here, to establish the fifth-order FLR among these fuzzified values, it is considered that A_{11} is caused by the previous five fuzzified values A_1, A_4, A_8, A_{16} and A_{13}. Hence, the fifth-order FLR is represented in the following form:

$$A_1, A_4, A_8, A_{16}, A_{13} \rightarrow A_{11} \tag{4.5.2}$$

Previously, most of the FTS models[1] use the current state's fuzzified value for defuzzification. The main downside of using such fuzzified value for defuzzification is that prediction scope of these models lie within the sample. For most of the real and complex problems, out of sample prediction (i.e., advance prediction) is very much essential. Therefore, in this model, the current state's fuzzified values are used to obtain the one step ahead forecasting results.

[1]References are: (Chen 2002; Chen and Chen 2011a, b; Chen et al. 2008; Tsai and Wu 2000).

Table 4.4 A sample of fifth-order FLRs for the June fuzzified daily temperature data set

Fifth-order FLR
$A_1, A_4, A_8, A_{16}, A_{13} \rightarrow ?\langle 6\rangle$
$A_4, A_8, A_{16}, A_{13}, A_{11} \rightarrow ?\langle 7\rangle$
$A_8, A_{16}, A_{13}, A_{11}, A_{12} \rightarrow ?\langle 8\rangle$
$A_{16}, A_{13}, A_{11}, A_{12}, A_{10} \rightarrow ?\langle 9\rangle$
...
$A_7, A_5, A_3, A_4, A_2 \rightarrow ?\langle 27\rangle$
$A_5, A_3, A_4, A_2, A_6 \rightarrow ?\langle 28\rangle$
$A_3, A_4, A_2, A_6, A_5 \rightarrow ?\langle 29\rangle$
$A_4, A_2, A_6, A_5, A_8 \rightarrow ?\langle 30\rangle$

The fifth-order FLRs obtained for the fuzzified daily temperature data are presented in Table 4.4. In this table, each symbol "?" represent the *desired output* for corresponding day "t" in the symbol "$\langle\rangle$", which would be determined by the proposed model.

Phase 4.5.5 Defuzzify the fuzzified time series data set.

In this model, we use the BPNN algorithm to defuzzify the fuzzified time series data set. The neural network architecture which is used here for defuzzification operation is presented in Sect. 4.2. The proposed model is based on the high-order FLRs, so to explain the defuzzification operation, we use the n*th*-order FLRs, where $n \geq 5$. The steps involved in the defuzzification operation are explained below.

Step 1. For forecasting day $Y(t)$, obtain the n*th*-order FLR, which can be represented in the following form:

$$A_{t(n-1)}, A_{t(n-2)}, \ldots, A_{t1}, A_{t0} \rightarrow ?\langle t\rangle, \tag{4.5.3}$$

where "t" represent a day which we want to forecast, and "n" is the order of FLR ($n \geq 5$). Here, $A_{t(n-1)}, A_{t(n-2)}, \ldots, A_{t2}, A_{t1}$ are the previous state's fuzzified values from days, $Y(t-n+1), Y(t-n+2) \ldots, Y(t-2)$, $Y(t-1)$.

Step 2. Find the intervals where the maximum membership values of the fuzzy sets $A_{t(n-1)}, A_{t(n-2)}, \ldots, A_{t1}, A_{t0}$ occur, and let these intervals be a_{n-1}, a_{n-2}, $\ldots, a_1, a_0$, respectively. All these intervals have the corresponding midpoints $C_{n-1}, C_{n-2}, \ldots, C_1, C_0$.

Table 4.5 A sample of advance prediction of the daily temperature for June

Day	Actual temperature	Predicted temperature
1	26.1	–
2	27.6	–
3	29	–
4	30.5	–
5	30	–
6	29.5	28.6
7	29.7	29.14
8	29.4	29.54
9	28.8	29.7
…	…	…
28	27.8	27.61
29	29	27.55
30	30.2	27.81

Step 3. Replace each previous state's fuzzified value of (4.5.3) with their corresponding mid-point as:

$$C_{n-1}, C_{n-2}, \ldots, C_1, C_0 \rightarrow ?\langle t\rangle, n \geq 5 \tag{4.5.4}$$

Step 4. Use the mid-points of (4.5.4) as inputs in the proposed BPNN architecture to compute the desired output "?" for the corresponding day "t".

The scaling of mid-points are done before beginning the defuzzification operation using min-max normalization (Han and Kamber 2001). For example, array of mid-points "X_i" are normalized based on the minimum and maximum values of "X_i". A mid-point "v" of "X_i" is normalized to "$\grave{v}$" by computing:

$$\grave{v} = \frac{v - min_A}{max_A - min_A}(new_{max_A} - new_{min_A}) + new_{min_A}, \tag{4.5.5}$$

where min_A and max_A are the minimum and maximum values of array "X_i" respectively. Min-max normalization maps a value "v" to "$\grave{v}$" in the range [new_{max_A}, new_{min_A}], where new_{max_A} represents "1" and new_{min_A} represents "0".

A sample of the results obtained for the June temperature data set are presented in Table 4.5. For rest of the months, similar approach is adopted for obtaining the results.

4.6 Experimental Results

Advance predicted values of temperature from June (1996) to September (1996) in Taipei for the fifth-order FLRs are presented in Table 4.6. The proposed model is also tested with different orders of FLRs. The performance of the model is evaluated with various statistical parameters, which are presented in Table 4.7. From Table 4.7, it is clear that mean of observed values are close to mean of predicted values. The comparison of SD values between observed and predicted values show that predictive skill of our proposed model is good for June, July and September. But, SD for predicted values for August shows slight deflection from SD of observed values. Forecasted results in terms of $RMSE$ indicate very small error rate. In Table 4.7, U values are closer to 0, which indicate the effectiveness of the proposed model.

During the training and testing processes of the neural network, a number of experiments were carried out to set additional parameters, *viz.*, initial weight, learning rate, epochs, learning radius and activation function to obtain the optimal results, and we have chosen the ones that exhibit the best behavior in terms of accuracy. The determined optimal values of all these parameters are given in Table 4.8.

Table 4.6 A sample of advance prediction of the daily temperature from June (1996) to September (1996) in Taipei (Unit: °C)

Day	Actual June	Predicted June	Actual July	Predicted July	Actual August	Predicted August	Actual September	Predicted September
1	26.1	–	29.9	–	27.1	–	27.5	–
2	27.6	–	28.4	–	28.9	–	26.8	–
3	29	–	29.2	–	28.9	–	26.4	–
4	30.5	–	29.4	–	29.3	–	27.5	–
5	30	–	29.9	–	28.8	–	26.6	–
6	29.5	28.6	29.6	29.25	28.7	28.37	28.2	26.81
7	29.7	29.14	30.1	29.19	29	28.82	29.2	26.97
8	29.4	29.54	29.3	29.59	28.2	28.69	29	27.5
9	28.8	29.7	28.1	29.45	27	28.43	30.3	27.69
...	...	...	...	...	...	...	...	...
28	27.8	27.61	28	29.46	27.7	28.33	24.2	25.81
29	29	27.55	29.3	28.86	26.2	28.23	23.3	25
30	30.2	27.81	27.9	28.49	26	27.99	23.5	25.05
31	–	–	26.9	28.37	27.7	27.32	–	–

Table 4.7 Performance analysis of the model for different orders of the FLRs

Order	Statistics	June	July	August	September
Fifth	$\bar{A}$ Observed (°C)	28.98	29.25	28.27	27.66
	$\bar{A}$ Predicted (°C)	28.91	29.32	28.29	27.77
	SD Observed (°C)	1.05	1.35	1.04	2.23
	SD Predicted (°C)	0.72	0.94	0.38	1.61
	RMSE (°C)	1.23	1.33	1.05	1.35
	U	0.0213	0.0225	0.0189	0.0189
Sixth	$\bar{A}$ Observed (°C)	28.95	29.24	28.26	27.64
	$\bar{A}$ Predicted (°C)	29.04	29.38	28.48	27.94
	SD Observed (°C)	1.07	1.38	1.04	2.27
	SD Predicted (°C)	0.65	0.83	0.39	1.61
	RMSE (°C)	1.27	1.36	1.03	1.43
	U	0.0219	0.0235	0.0185	0.0256
Seventh	$\bar{A}$ Observed (°C)	28.92	29.20	28.23	27.57
	$\bar{A}$ Predicted (°C)	29.05	29.34	28.43	27.97
	SD Observed (°C)	1.08	1.40	1.05	2.30
	SD Predicted (°C)	0.53	0.74	0.35	1.57
	RMSE (°C)	1.22	1.37	1.02	1.39
	U	0.0210	0.0239	0.0180	0.0249
Eighth	$\bar{A}$ Observed (°C)	28.90	29.20	28.23	27.51
	$\bar{A}$ Predicted (°C)	29.04	29.27	28.44	28.01
	SD Observed (°C)	1.10	1.43	1.07	2.33
	SD Predicted (°C)	0.42	0.72	0.33	1.44
	RMSE (°C)	1.25	1.47	1.03	1.57
	U	0.0216	0.0257	0.0185	0.0282

Table 4.8 Additional parameters and their values during the training and testing processes of neural network

Additional parameter	Input value
Initial weight	0.3
Learning rate	0.5
Epochs	10000
Learning radius	3
Activation function	Sigmoid

4.7 Advance Prediction of BSE

BSE Limited formerly known as Bombay Stock Exchange (BSE) is a stock exchange located in Mumbai (India) and is the oldest stock exchange in Asia. The equity market capitalization of the companies listed on the BSE was US$1 trillion as of December 2011, making it the 6th largest stock exchange in Asia and the 14th largest in the world (www.World-exchanges.org). The BSE has the largest number of listed companies in the world.

To further demonstrate the applicability of the proposed model, daily stock exchange price of the BSE is tried to be predicted. The BSE data set for the period 7/30/2012 to 9/11/2012 is collected from: http://in.finance.yahoo.com/.

The predicted values of the BSE based on the fifth-order FLRs are presented in Table 4.9. To check the efficiency of the model, results are also obtained with different

Table 4.9 A sample of advance prediction of the BSE price from 7/30/2012 to 9/11/2012 (In Rupee)

Date (mm/dd/yy)	Actual price	Predicted price
7/30/2012	17143.68	–
7/31/2012	17236.18	–
8/1/2012	17257.38	–
8/2/2012	17224.36	–
8/3/2012	17197.93	–
8/6/2012	17412.96	17177.27
8/7/2012	17601.78	17259.28
8/8/2012	17600.56	17299.01
...	...	...
9/7/2012	17683.73	17374.28
9/10/2012	17766.78	17369.4
9/11/2012	17852.95	17469.24

Table 4.10 Performance analysis of advance prediction of the BSE price for different orders of FLRs

Statistics	Fifth -order	Sixth -order	Seventh -order	Eight -order
$\bar{A}$ Observed (Rupee)	17612.86	17621.19	17622.03	17623.01
$\bar{A}$ Predicted (Rupee)	17523.59	17538.02	17550.14	17561.56
SD Observed (Rupee)	169.51	167.84	171.56	175.54
SD Predicted (Rupee)	165.76	152.44	143.56	135.84
RMSE (Rupee)	203.10	201.63	193.19	186.77
U	0.0058	0.0057	0.0055	0.0053

orders of FLRs. The performance of the model is evaluated with various statistical parameters, which are presented in Table 4.10. All these statistical analyzes signify the robustness of the proposed model for advance prediction of the BSE price.

4.8 Discussion

This chapter presents a novel approach combining ANN with FTS for building a time series forecasting expert system. For training process, the daily average temperature data of Taipei from June 1, 1996 to June 30, 1996 are used; while for testing process, the daily average temperature data of Taipei from July, 1996 to September, 1996 are considered. The proposed model is also validated by predicting the BSE price from the period 7/30/2012 to 9/11/2012.

In this work, we have incorporated "RPD" approach for determining the lengths of the intervals effectively, which is an improvement over the original works presented by Song and Chissom (1993a, b, 1994). Also, many existing FTS models as discussed earlier, use the current state's ($Y_{t-1} \rightarrow Y_t$) fuzzified values for defuzzification, and limit their predictive skill within the sample. So, to make the prediction out-of-sample, we have used one advanced state's ($Y_t \rightarrow Y_{t+1}$) fuzzified values for defuzzification. In this study, for defuzzification operation, an ANN based architecture is developed, which is based on the BPNN algorithm. The proposed neural network architecture takes the previous state's fuzzified values as inputs and outputs are computed in advance.

In this study, we try to obtain the forecasting results with optimal number of intervals. To obtain the results for the months – June, July, August and September, only 17, 17, 20 and 21 intervals are used respectively. On the other hand, for the BSE price prediction, only 21 intervals are employed. The performance of the model is evaluated with different orders of fuzzy logical relations, which signify the efficiency of the proposed model in case of temperature as well as stock exchange price prediction.

References

Aladag CH, Basaran MA, Egrioglu E, Yolcu U, Uslu VR (2009) Forecasting in high order fuzzy times series by using neural networks to define fuzzy relations. Expert Syst Appl 36(3):4228–4231

Chang PC, Liu CH, Fan CY (2009) Data clustering and fuzzy neural network for sales forecasting: a case study in printed circuit board industry. Knowl-Based Syst 22(5):344–355

Chen SM (2002) Forecasting enrollments based on high-order fuzzy time series. Cybern Syst Int J 33(1):1–16

Chen SM, Chen CD (2011a) Handling forecasting problems based on high-order fuzzy logical relationships. Expert Syst Appl 38(4):3857–3864

Chen SM, Chen CD (2011b) Handling forecasting problems based on high-order fuzzy logical relationships. Expert Syst Appl 38(4):3857–3864

Chen SM, Hwang JR (2000) Temperature prediction using fuzzy time series. IEEE Trans Syst Man Cybern Part B Cybern 30:263–275

Chen TL, Cheng CH, Teoh HJ (2008) High-order fuzzy time-series based on multi-period adaptation model for forecasting stock markets. Phys A Stat Mech Appl 387(4):876–888

Cheng CH, Chen TL, Wei LY (2010) A hybrid model based on rough sets theory and genetic algorithms for stock price forecasting. Inf Sci 180(9):1610–1629

Hadavandi E, Shavandi H, Ghanbari A (2010) Integration of genetic fuzzy systems and artificial neural networks for stock price forecasting. Knowl-Based Syst 23(8):800–808

Han J, Kamber M (2001) Data Mining Concepts Tech, 1st edn. Morgan Kaufmann Publishers, USA

Huang YL, Horng SJ, He M, Fan P, Kao TW, Khan MK, Lai JL, Kuo IH (2011) A hybrid forecasting model for enrollments based on aggregated fuzzy time series and particle swarm optimization. Expert Syst Appl 38(7):8014–8023

Huarng K, Yu THK (2006) The application of neural networks to forecast fuzzy time series. Phys A Stat Mech Appl 363(2):481–491

Keles A, Kolcak M, Keles A (2008) The adaptive neuro-fuzzy model for forecasting the domestic debt. Knowl-Based Syst 21(8):951–957

Kuo IH, Horng SJ, Kao TW, Lin TL, Lee CL, Pan Y (2009) An improved method for forecasting enrollments based on fuzzy time series and particle swarm optimization. Expert Syst Appl 36(3, Part 2):6108–6117

Kuo IH, Horng SJ, Chen YH, Run RS, Kao TW, Chen RJ, Lai JL, Lin TL (2010) Forecasting TAIFEX based on fuzzy time series and particle swarm optimization. Expert Syst Appl 37(2):1494–1502

Pal SK, Mitra P (2004) Case generation using rough sets with fuzzy representation. IEEE Trans Knowl Data Eng 16(3):292–300

Roy S, Chakraborty U (2013) Introduction to soft computing. Dorling Kindersley (India) Pvt Ltd, New Delhi

Song Q, Chissom BS (1993a) Forecasting enrollments with fuzzy time series—Part I. Fuzzy Sets Syst 54(1):1–9

Song Q, Chissom BS (1993b) Fuzzy time series and its models. Fuzzy Sets Syst 54(1):1–9

Song Q, Chissom BS (1994) Forecasting enrollments with fuzzy time series—Part II. Fuzzy Sets Syst 62(1):1–8

Sturges H (1926) The choice of a class-interval. J Am Stat Assoc 21:65–66

Teoh HJ, Cheng CH, Chu HH, Chen JS (2008) Fuzzy time series model based on probabilistic approach and rough set rule induction for empirical research in stock markets. Data Knowl Eng 67(1):103–117

Tsai CC, Wu SJ (2000) Forecasting enrolments with high-order fuzzy time series. In: 19th international conference of the North American. Fuzzy Information Processing Society, Atlanta, GA, pp 196–200

Wilson ID, Paris SD, Ware JA, Jenkins DH (2002) Residential property price time series forecasting with neural networks. Knowl-Based Syst 15(5–6):335–341

Yang L, Dawson CW, Brown MR, Gell M (2006) Neural network and GA approaches for dwelling fire occurrence prediction. Knowl-Based Syst 19(4):213–219

Yu THK, Huarng KH (2008) A bivariate fuzzy time series model to forecast the TAIEX. Expert Syst Appl 34(4):2945–2952

Chapter 5
Two-Factors High-Order Neuro-Fuzzy Forecasting Model

The essence of mathematics is not to make simple things complicated, but to make complicated things simple

By S. Gudden

Abstract FTS forecasting method has been applied in several domains, such as stock market price, temperature, sales, crop production, and academic enrollments. In this chapter, we introduce a model to deal with forecasting problems of two-factors. The proposed model is designed using FTS and ANN. In a FTS, the length of intervals in the universe of discourse always affects the results of forecasting. Therefore, an ANN based technique is employed for determining the intervals of the historical time series data sets by clustering them into different groups. The historical time series data sets are then fuzzified, and the high-order FLRs are established among fuzzified values based on FTS method. The chapter also introduces some rules for interval weighing to defuzzify the fuzzified time series data sets. From experimental results, it is observed that the proposed model exhibits higher accuracy than those of existing two-factors FTS models.

Keywords FTS · Two-factors · Temperature · FLR · ANN

5.1 Background and Related Literature

Chen and Hwang (2000) forecasted the daily average temperature of Taipei based on two-factors FTS. In this model, first factor is daily temperature, whereas the second factor is daily cloud density. They proposed two algorithms—Algorithm-B and Algorithm-B*. Their experimental results show that the accuracy rate of Algorithm-B* is better than the Algorithm-B. Lee et al. (2006) proposed a new method to forecast the daily average temperature of Taipei and TAIFEX. In this model, high order FLR is constructed to increase the forecasting accuracy. Chang and Chen (2009) forecasted the daily temperature using FCM and fuzzy rules inter-

P. Singh, *Applications of Soft Computing in Time Series Forecasting*,
Studies in Fuzziness and Soft Computing 330, DOI 10.1007/978-3-319-26293-2_5

Algorithm 3 SOFM Algorithm

Step 1: Initialize the weights (W_{uv}) and learning rate (α).
Step 2: When stopping condition is false, then perform Steps 2–8.
Step 3: For each input vector (X), perform Steps 3–5.
Step 4: For each $v = 1$ to m, compute the square of the Euclidean distance as:

$$D(v) = \sum_{u=1}^{n} (X_u - W_{uv})^2 \tag{5.1.1}$$

Step 5: Obtain winning unit index (J), so that $D(J) =$ minimum.
Step 6: Calculate weights of winning unit as:

$$W_{uv}(new) = W_{uv}(old) + \alpha[X_u - W_{uv}(old)] \tag{5.1.2}$$

Step 7: Reduce the learning rate (α) by using the following formula:

$$\alpha(t+1) = 0.5\alpha(t) \tag{5.1.3}$$

Step 8: Reduce radius of topological neighborhood network.
Step 9: Test for stopping condition of the network.

polation techniques. In this model, rules are constructed based on the FCM clustering algorithm. Then, this model performs fuzzy inference based on the multiple fuzzy rules interpolation scheme. Based on two-factors high-order FTS and automatic clustering techniques, Wang and Chen (2009) proposed a new method to predict the daily average temperature and TAIFEX. Lee et al. (2007, 2008) presented a new method for temperature prediction and the TAIFEX forecasting based on two-factors high-orders FLRs by hybridizing GA with FTS method.

5.2 Clustering Using Self-organizing Feature Maps ANN

Data clustering is a popular approach for automatically finding groups in multidimensional data. Self-organizing feature maps (SOFM), which is a class of neural networks developed by Kohonen (1990), can be utilized for clustering a data set. The SOFM is trained by an iterative unsupervised or self-organizing procedure (Liao 2005). It converts the patterns of arbitrary dimensionality into response of one-dimensional or two-dimensional arrays of neurons, i.e., it converts a wide pattern space into feature map. The training process of SOFM is presented in Algorithm 3 (Sivanandam and Deepa 2007). Based on the clusters obtained by applying SOFM, the historical time series data sets can be partitioned into different length of intervals.

5.3 Hybridized Model for Two-Factors Time Series Data

In this section, we introduce a new forecasting model based on hybridization of ANN with FTS. The proposed model consists of six phases as presented below. For verification of model, the historical data sets of the daily average temperature and the daily cloud density from June, 1996 to September, 1996 in Taipei, Taiwan (Chen and Hwang 2000) are used, which are shown in Tables 5.1 and 5.2, respectively. In these data sets, the daily average temperature is called the main-factor, and the daily average cloud density is called the second-factor.

In the following, we apply the proposed model to predict the daily temperature of Taipei from June, 1996 to September, 1996. To explain the functionality of each phase of the model, the daily average temperature and the daily cloud density data sets from June 1, 1996 to June 30, 1996 are considered as an example. Each phase of the model is explained below.

Phase 5.3.1 Divide the universe of discourse into different lengths of intervals, and compute weights for each interval.

[Explanation] Define the universe of discourse "A" of the main-factor and the universe of discourse "B" of the second-factor of the historical time series data sets. Let $A = [M_{min}, M_{max}]$, where M_{min} and M_{max} are the minimum and maximum values of the main-factor respectively. Let $B = [N_{min}, N_{max}]$, where N_{min} and N_{max} are the minimum and maximum values of the second-factor respectively.

Based on Tables 5.1 and 5.2, we have the universe of discourse of the daily average temperature $A = [26.1, 30.9]$, and the universe of discourse of the cloud density

Table 5.1 A sample of historical data of the daily average temperature from June, 1996 to September, 1996 in Taipei (Chen and Hwang 2000) (Unit: °C)

Day	Month			
	June	July	August	September
1	26.1	29.9	27.1	27.5
2	27.6	28.4	28.9	26.8
3	29.0	29.2	28.9	26.4
4	30.5	29.4	29.3	27.5
...	...	...	...	...
25	27.7	28.9	28.6	25.8
26	27.1	28.0	28.7	26.4
27	28.4	28.6	29.0	25.6
28	27.8	28.0	27.7	24.2
29	29.0	29.3	26.2	23.3
30	30.2	27.9	26.0	23.5
31	–	26.9	27.7	–

Table 5.2 A sample of historical data of the daily average cloud density from June, 1996 to September, 1996 in Taipei (Chen and Hwang 2000) (Unit: %)

Day	Month			
	June	July	August	September
1	36	15	100	29
2	23	31	78	53
3	23	26	68	66
4	10	34	44	50
…	…	…	…	…
25	14	100	29	40
26	25	100	31	30
27	29	91	41	34
28	55	84	14	59
29	29	38	28	83
30	19	46	33	38
31	–	95	26	–

Table 5.3 A sample of intervals, corresponding data, centroids and weights for the daily temperature for June, 1996

Interval	Corresponding data	Centroid	Weight
$a_1 = [26.1,\ 26.1)$	(26.1)	26.1	1
$a_2 = [27.1,\ 27.1)$	(27.1)	27.1	1
$a_3 = [27.4,\ 27.4)$	(27.4)	27.4	1
$a_4 = [27.5,\ 27.6)$	(27.5, 27.6)	27.55	2
…	…	…	…
$a_{10} = [29.3,\ 29.4)$	(29.3, 29.4, 29.4)	29.37	3
$a_{11} = [29.5,\ 29.7)$	(29.5, 29.5, 29.7)	29.57	3
$a_{12} = [30.0,\ 30.3)$	(30, 30.2, 30.2, 30.3)	30.18	4
$a_{13} = [30.5,\ 30.9)$	(30.5, 30.8, 30.9)	30.73	3

$B = [10, 96]$. By applying the SOFM algorithm, divide the universe of discourse "A" into different lengths of intervals as $a_1, a_2, \ldots,$ and a_n. Similarly, divide the universe of discourse "B" into different lengths of intervals as $b_1, b_2, \ldots,$ and b_n. For each interval, the centroid is calculated by taking the mean of the upper bound and lower bound of the interval. Each interval bears a weight equal to the frequency of the interval. The resulting intervals, centroids and weights for the considered data sets are shown in Tables 5.3 and 5.4.

Phase 5.3.2 Define linguistic terms for each of the interval.

Table 5.4 A sample of intervals, corresponding data, centroids and weights for the daily cloud density for June, 1996

Interval	Corresponding data	Centroid	Weight
$b_1 = [10,\ 10)$	(10)	10	1
$b_2 = [13,\ 14)$	(13, 14)	13.5	2
$b_3 = [15,\ 15)$	(15)	15	1
$b_4 = [19,\ 19)$	(19, 19)	19	2
...	...	...	...
$b_{12} = [43,\ 45)$	(44, 45)	44	2
$b_{13} = [46,\ 46)$	(46)	46	1
$b_{14} = [55,\ 56)$	(55, 55, 56, 60)	56.5	4
$b_{15} = [63,\ 63)$	(63)	63	1
$b_{16} = [96,\ 96)$	(96)	96	1

[Explanation] The universe of discourse "A" of the main-factor is divided into n intervals (*i.e.*, $a_1, a_2, \ldots,$ and a_n). Assume that there are n linguistic variables (*i.e.*, $U_1, U_2, \ldots, U_n$) represented by fuzzy sets, where $1 \le i \le n$, shown as follows:

$$\begin{aligned}
U_1 &= 1/a_1 + 0.5/a_2 + 0/a_3 + \ldots + 0/a_{n-2} + 0/a_{n-1} + 0/a_n,\\
U_2 &= 0.5/a_1 + 1/a_2 + 0.5/a_3 + \ldots + 0/a_{n-2} + 0/a_{n-1} + 0/a_n,\\
U_3 &= 0/a_1 + 0.5/a_2 + 1/a_3 + \ldots + 0/a_{n-2} + 0/a_{n-1} + 0/a_n,\\
&\vdots\\
U_n &= 0/a_1 + 0/a_2 + 0/a_3 + \ldots + 0/a_{n-2} + 0.5/a_{n-1} + 1/a_n.
\end{aligned}$$

Similarly, the universe of discourse "B" of the second-factor is divided into m intervals (*i.e.*, $b_1, b_2, \ldots,$ and b_m). Assume that there are m linguistic variables (*i.e.*, $V_1, V_2, \ldots, V_m$) represented by fuzzy sets, where $1 \le i \le m$, shown as follows:

$$\begin{aligned}
V_1 &= 1/b_1 + 0.5/b_2 + 0/b_3 + \ldots + 0/b_{m-2} + 0/b_{m-1} + 0/b_m,\\
V_2 &= 0.5/b_1 + 1/b_2 + 0.5/b_3 + \ldots + 0/b_{m-2} + 0/b_{m-1} + 0/b_m,\\
V_3 &= 0/b_1 + 0.5/b_2 + 1/b_3 + \ldots + 0/b_{m-2} + 0/b_{m-1} + 0/b_m,\\
&\vdots\\
V_m &= 0/b_1 + 0/b_2 + 0/b_3 + \ldots + 0/b_{m-2} + 0.5/b_{m-1} + 1/b_m.
\end{aligned}$$

The maximum membership values of both U_i and V_i occur at intervals a_i and b_i respectively.

Phase 5.3.3 Fuzzify the historical time series data sets of the main-factor and the second-factor.

Table 5.5 A sample of fuzzified daily temperature (with their corresponding centroids and weights) and cloud density for June, 1996

Day	Temperature	Fuzzified temperature	Centroid	Weight	Cloud density	Fuzzified cloud density
1	26.1	U_1	26.1	1	36	V_{11}
2	27.6	U_4	27.55	2	23	V_6
3	29.0	U_9	29.0	3	23	V_6
4	30.5	U_{13}	30.73	3	10	V_1
...	...	...	...	...	...	...
28	27.8	U_6	27.8	2	55	V_{14}
29	29.0	U_9	29.0	3	29	V_8
30	30.2	U_{12}	30.18	4	19	V_4

[Explanation] If the time series data of the main-factor belongs to the interval a_i, where $1 \leq i \leq n$, then fuzzify the time series data of the main-factor into fuzzy set U_i. Similarly, if the time series data of the second-factor belongs to the interval b_i, where $1 \leq i \leq m$, then fuzzify the time series data of the second-factor into fuzzy set V_i.

The fuzzified values of the main-factor and second-factor for June, 1996 time series data sets are shown in Table 5.5. The fourth and fifth columns of Table 5.5 represent the centroids and weights of the corresponding intervals for the main-factor, respectively. In Table 5.5, only fuzzified values of the second-factor are shown (last column), because for forecasting the main-factor, only fuzzified values of the second-factor are required.

Phase 5.3.4 Establish the FLRs between the fuzzified main-factor and the fuzzified second-factor.

[Explanation] We can establish the nth-order FLRs based on the fuzzified main-factor and the fuzzified second-factor. If there exists a FLR between U_i and V_i, where U_i and V_i denote the fuzzified main-factor and second-factor of day "i" respectively, then the two-factors nth-order FLR can be represented as follows:

$$((U_{ni}, V_{ni}), \ldots, (U_{n2}, V_{n2}), (U_{n1}, V_{n1})) \rightarrow U_i \tag{5.3.1}$$

Here, (U_{ni}, V_{ni}), ..., (U_{n2}, V_{n2}), (U_{n1}, V_{n1}) and U_i represent fuzzified values of day $n-i$, ..., day $n-2$, day $n-1$, and day i respectively, where $2 \leq i \leq n$. The left-hand side and right-hand side of FLR (5.3.1) are called the *previous state* and the *current state*, respectively. Here, $U_{ni}, \ldots$, U_{n2}, and U_{n1} represent the fuzzified values of the main-factor of days $n-i, \ldots$, $n-2$, and $n-1$, respectively. Similarly, V_{ni}, ..., V_{n2}, and V_{n1} represent the fuzzified values of the second-factor of days $n-i$, ..., $n-2$, and $n-1$, respectively.

Table 5.6 A sample of two-factors first-order FLRs between the fuzzified temperature and cloud density data of June, 1996

Two-factors first-order FLRs
$(U_1, V_{11}) \rightarrow U_4$
$(U_4, V_6) \rightarrow U_9$
$(U_9, V_6) \rightarrow U_{13}$
$(U_{13}, V_1) \rightarrow U_{12}$
...
$(U_6, V_{14}) \rightarrow U_9$
$(U_9, V_8) \rightarrow U_{12}$
$(U_{12}, V_4) \rightarrow ?$

Table 5.7 A sample of two-factors second-order FLRs between the fuzzified temperature and cloud density data of June, 1996

Two-factors second-order FLRs
$((U_1, V_{11}), (U_4, V_6)) \rightarrow U_9$
$((U_4, V_6), (U_9, V_6)) \rightarrow U_{13}$
$((U_9, V_6), (U_{13}, V_1)) \rightarrow U_{12}$
$((U_{13}, V_1), (U_{12}, V_2)) \rightarrow U_{11}$
...
$((U_7, V_8), (U_6, V_{14})) \rightarrow U_9$
$((U_6, V_{14}), (U_9, V_8)) \rightarrow U_{12}$
$((U_9, V_8), (U_{12}, V_4)) \rightarrow ?$

Based on FLR (5.3.1) and Table 5.5, the first-order and the second-order FLRs of two-factors are formed, which are shown in Tables 5.6 and 5.7, respectively. In Tables 5.6 and 5.7, the symbol "?" represents an unknown value.

Phase 5.3.5 Form the FLRGs.

[Explanation] If the nth-order FLRs have the same previous state, then the nth-order FLRs can be divided into a nth-order FLRG. Consider the following nth-order FLRs given as follows:

$$((U_{ni}, V_{ni}), \ldots, (U_{n2}, V_{n2}), (U_{n1}, V_{n1})) \rightarrow U_k$$
$$((U_{ni}, V_{ni}), \ldots, (U_{n2}, V_{n2}), (U_{n1}, V_{n1})) \rightarrow U_s$$
$$\vdots$$
$$((U_{ni}, V_{ni}), \ldots, (U_{n2}, V_{n2}), (U_{n1}, V_{n1})) \rightarrow U_n$$

Then, the nth-order FLRG can be formed as follows:

$$((U_{ni}, V_{ni}), \ldots, (U_{n2}, V_{n2}), (U_{n1}, V_{n1})) \rightarrow U_k, U_s, \ldots, U_n \tag{5.3.2}$$

Table 5.8 A sample of two-factors first-order FLRGs of the fuzzified temperature and cloud density data of June, 1996

Two-factors first-order FLRGs
Group 1: $(U_1, V_{11}) \rightarrow U_4$
Group 2: $(U_4, V_6) \rightarrow U_9$
Group 3: $(U_9, V_6) \rightarrow U_{13}$
Group 4: $(U_{13}, V_1) \rightarrow U_{12}$
...
Group 27: $(U_6, V_{14}) \rightarrow U_9$
Group 28: $(U_9, V_8) \rightarrow U_{12}$
Group 29: $(U_{12}, V_4) \rightarrow ?$

Table 5.9 A sample of two-factors second-order FLRGs of the fuzzified temperature and cloud density data of June, 1996

Two-factors second-order FLRGs
Group 1: $((U_1, V_{11}), (U_4, V_6)) \rightarrow U_9$
Group 2: $((U_4, V_6), (U_9, V_6)) \rightarrow U_{13}$
Group 3: $((U_9, V_6), (U_{13}, V_1)) \rightarrow U_{12}$
Group 4: $((U_{13}, V_1), (U_{12}, V_2)) \rightarrow U_{11}$
...
Group 27: $((U_7, V_8), (U_6, V_{14})) \rightarrow U_9$
Group 28: $((U_6, V_{14}), (U_9, V_8)) \rightarrow U_{12}$
Group 29: $((U_9, V_8), (U_{12}, V_4)) \rightarrow ?$

The first-order FLRGs are formed based on Table 5.6, which are shown in Table 5.8; and the second-order FLRGs are formed based on Table 5.7, which are shown in Table 5.9. If the same FLR appears more than once, it is included only once in the group.

Phase 5.3.6 Compute the forecasted values.

[Explanation] To compute the forecasted values, the rules for interval weighing are proposed. These rules are presented as follows:

Rule 1. For forecasting day, $D(t)$, the previous state's fuzzified values of the main-factor and the second-factor are considered from days, $D(t-n), \ldots, D(t-2)$ to $D(t-1)$; where "t" is the current day which we want to forecast, and "n" is the order of FLRs. The **Rule 1** is applicable only if there is only one fuzzified value in the current state. The steps under **Rule 1** are given as follows:

Step 1. For forecasting day, $D(t)$, obtain the previous state's fuzzified values of the main-factor and the second-factor from days $D(t-n)$ to $D(t-1)$ as $(U_{ni}, V_{ni}), \ldots, (U_{n2}, V_{n2})$ and (U_{n1}, V_{n1}).

Step 2. Find the FLRG whose previous state is $((U_{ni}, V_{ni}), \ldots, (U_{n2}, V_{n2}), (U_{n1}, V_{n1}))$, and current state is U_k, *i.e.*, the FLRG is in the form of

$((U_{ni}, V_{ni}), \ldots, (U_{n2}, V_{n2}), (U_{n1}, V_{n1})) \rightarrow U_k$, then the forecasted value is calculated based on the following step.

Step 3. Find the interval where the maximum membership value of U_k occurs. Let this interval be a_k. This interval a_k has the corresponding centroid C_k. This centroid C_k is the forecasted value for day, $D(t)$.

Rule 2. This rule is applicable if there are more than one fuzzified values in the current state. The steps under **Rule 2** are given as follows:

Step 1. For forecasting day, $D(t)$, obtain the previous state's fuzzified values of the main-factor and the second-factor from days $D(t-n)$ to $D(t-1)$ as $(U_{ni}, V_{ni}), \ldots, (U_{n2}, V_{n2})$ and (U_{n1}, V_{n1}).

Step 2. Find the FLRG whose previous state is $((U_{ni}, V_{ni}), \ldots, (U_{n2}, V_{n2}), (U_{n1}, V_{n1}))$, and current state is $U_k, U_s, \ldots, U_n$, *i.e.*, the FLRG is in the form of $((U_{ni}, V_{ni}), \ldots, (U_{n2}, V_{n2}), (U_{n1}, V_{n1})) \rightarrow U_k, U_s, \ldots, U_n$, then the forecasted value is calculated based on the following step.

Step 3. Find the intervals where the maximum membership values of $U_k, U_s, \ldots, U_n$ occur, and let these intervals be $a_k, a_s, \ldots, a_n$, respectively. These intervals have the corresponding centroids $C_k, C_s, \ldots, C_n$ and weights $W_k, W_s, \ldots, W_n$, respectively.

Step 4. The forecasted value for day, $D(t)$ is calculated as follows:

$$Forecast\ (t) = \frac{C_k W_k + C_s W_s + \ldots + C_n W_n}{W_k + W_s + \ldots + W_n} \tag{5.3.3}$$

Rule 3. This rule is applicable only if there is an unknown value in the current state. The steps under **Rule 3** are given as follows:

Step 1. For forecasting day, $D(t)$, obtain the previous state's fuzzified values of the main-factor and the second-factor from days $D(t-n)$ to $D(t-1)$ as $(U_{ni}, V_{ni}), \ldots, (U_{n2}, V_{n2})$ and (U_{n1}, V_{n1}).

Step 2. Find the FLRG whose previous state is $((U_{ni}, V_{ni}), \ldots, (U_{n2}, V_{n2}), (U_{n1}, V_{n1}))$, and current state is "?"(the symbol "?" represents an unknown value), *i.e.*, the FLRG is in the form of $((U_{ni}, V_{ni}), \ldots, (U_{n2}, V_{n2}), (U_{n1}, V_{n1})) \rightarrow ?$, then the forecasted value is calculated based on the following step.

Step 3. Find the intervals where the maximum membership values of $U_{ni}, \ldots, U_{n2}, U_{n1}$ occur, and let these intervals be $a_{n-i}, \ldots, a_{n-2}, a_{n-1}$, respectively. These intervals have the corresponding centroids $C_{n-i}, \ldots, C_{n-2}, C_{n-1}$ and weights $W_{n-i}, \ldots, W_{n-2}, W_{n-1}$, respectively.

Step 4. The forecasted value for day, $D(t)$ is calculated as follows:

$$Forecast\ (t) = \frac{C_{n-i} W_{n-i} + \ldots + C_{n-2} W_{n-2} + C_{n-1} W_{n-1}}{W_{n-i} + \ldots + W_{n-2} + W_{n-1}} \tag{5.3.4}$$

The daily average temperatures of June, 1996 are forecasted based on the two-factors second-order fuzzy logical time series, which is shown in Table 5.10.

Table 5.10 A sample of forecasted daily average temperature of June, 1996 based on the two-factors second-order fuzzy logical time series (Unit: °C)

Day	Actual temperature	Actual cloud density	Forecasted temperature
1	26.1	36	–
2	27.6	23	–
3	29.0	23	29.00
4	30.5	10	30.73
5	30.0	13	30.18
6	29.5	30	29.57
...	...	...	...
28	27.8	55	27.80
29	29.0	29	29.00
30	30.2	19	30.18

Based on the proposed method, we have presented here two examples to compute forecasted values of daily average temperature as follows:

[Example 1] Based on two-factors first-order fuzzy logical time series, an example is presented here to forecast the temperature on day, $D(t)$. Suppose, we want to forecast the temperature on June 7, 1996, in Taipei. To compute this value, the fuzzified temperature and cloud density values of the previous state are required. For forecasting day, *D*(June 7), the fuzzified temperature and cloud density values for day, *D*(June 6) are obtained from Table 5.5, which are U_{11} and V_9, respectively. Then, obtain the FLRG whose previous state is (U_{11}, V_9) from Table 5.8. In this case, the FLRG is $(U_{11}, V_9) \rightarrow U_{11}, U_8$ (*i.e.*, Group 6). Therefore, **Rule 2** is applicable here, because the current state has two fuzzified values. Now, find the intervals where the maximum membership values of U_{11} and U_8 occur from Table 5.3, which are a_{11} and a_8, respectively. The corresponding centroid and weight for the interval a_{11} are 29.57 and 3 respectively. The corresponding centroid and weight for the interval a_8 are 28.75 and 4 respectively. Now, based on Eq. 5.3.3, the forecasted temperature for day, *D*(June 7) can be computed as:

$$\frac{(29.57 \times 3 + 28.75 \times 4)}{3+4} = 29.10$$

[Example 2] Based on two-factors second-order fuzzy logical time series, an example is presented here to forecast the temperature on day, $D(t)$. Suppose, we want to forecast the temperature on June 4, 1996, in Taipei. To compute this value, the fuzzified temperature and cloud density values of the previous state are required. For forecasting day, *D*(June 4), the fuzzified temperature and cloud density values for days, *D*(June 2) and *D*(June 3) are obtained from Table 5.5, which are (U_4, V_6) and (U_9, V_6), respectively. Then, obtain the FLRG whose previous state is $((U_4, V_6), (U_9, V_6))$ from Table 5.9. In this case, the FLRG is $((U_4, V_6), (U_9, V_6)) \rightarrow U_{13}$ (*i.e.*, Group 2). Therefore, **Rule 1** is applicable here, because in the current state, only one fuzzified value is available. Now, find the interval where the maximum membership value for fuzzy set U_{13} occurs from Table 5.3, which is a_{13}. The interval a_{13} has the centroid 30.73, which is the forecasted temperature for day, D(June 4).

5.4 Experimental Results

The proposed model computes the forecasted values with the help of hybridization of ANN (SOFM neural network) with the FTS. For training process, the daily temperature and the daily cloud density data sets from June 1, 1996 to June 30, 1996 are employed. In the testing process, the data sets of the daily temperature and the daily cloud density from July, 1996 to September, 1996 are used.

During the learning process of neural network, different experiments were carried out to set additional parameters like learning rate, epochs, initial weight, learning radius, etc. to obtain optimal results, and we have chosen the ones that exhibit the best behavior in terms of accuracy. The determined optimal values of all these parameters are listed in Table 5.11.

The main downside of FTS forecasting model is that increase in the number of intervals increases accuracy rate of forecasting, and decreases the fuzziness of time series data sets. Therefore, in this study, the parameter called "optimum number of intervals" for the main-factor and second-factor time series data sets are decided using a heuristic approach. We have tried different values for this parameter, and calcul ate the AFER for different orders for the months: June, July, August and September.

All these experimental results are plotted in graphs for different orders and intervals as shown in Fig. 5.1, and we have chosen the "optimum number of intervals"

Table 5.11 Additional parameters and their values during the learning process of SOFM neural network

Serial number	List of additional parameter	Value
1	Learning rate	0.5
2	Epochs	100
3	Initial weight	0.3
4	Learning radius	3
5	Optimum number of intervals for main-factor (June)	13
	Optimum number of intervals for second-factor (June)	15
6	Optimum number of intervals for main-factor (July)	12
	Optimum number of intervals for second-factor (July)	15
7	Optimum number of intervals for main-factor (August)	16
	Optimum number of intervals for second-factor (August)	15
8	Optimum number of intervals for main-factor (September)	15
	Optimum number of intervals for second-factor (September)	15

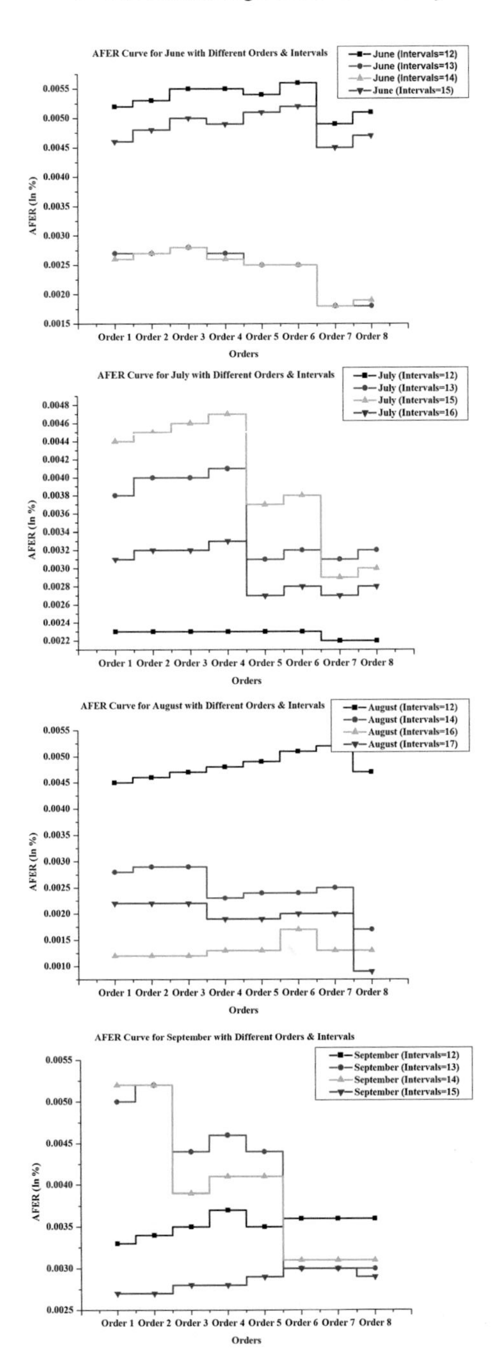

Fig. 5.1 AFER curves for June, July, August and September (*top* to *bottom*) with different orders and intervals

Table 5.12 The AFERs to forecast the temperature from June, 1996 to September, 1996 in Taipei

Model	Month	Order								
		First	Second	Third	Fourth	Fifth	Sixth	Seventh	Eighth	
Proposed	June	0.27 %	0.27 %	0.28 %	0.27 %	0.25 %	0.25 %	0.18 %	0.18 %	
	July	0.23 %	0.23 %	0.23 %	0.23 %	0.23 %	0.23 %	0.22 %	0.22 %	
	August	0.12 %	0.12 %	0.12 %	0.13 %	0.13 %	0.17 %	0.13 %	0.13 %	
	September	0.27 %	0.27 %	0.28 %	0.28 %	0.29 %	0.30 %	0.30 %	0.29 %	
	Month	Window bases								
		w = 2	w = 3	w = 4	w = 5	w = 6	w = 7	w = 8		
Chen and Hwang (2000)	June	2.88 %	3.16 %	3.24 %	3.33 %	3.39 %	3.53 %	3.67 %		
	July	3.04 %	3.76 %	4.08 %	4.17 %	4.35 %	4.38 %	4.56 %		
	August	2.75 %	2.77 %	3.30 %	3.40 %	3.18 %	3.15 %	3.19 %		
	September	3.29 %	3.10 %	3.19 %	3.22 %	3.39 %	3.38 %	3.29 %		
	Month	Order								
		First	Second	Third	Fourth	Fifth	Sixth	Seventh	Eighth	
Lee (2006)	June	1.44 %	0.80 %	0.76 %	0.79 %	0.76 %	0.79 %	0.79 %	0.81 %	
	July	1.59 %	0.96 %	0.96 %	0.98 %	0.97 %	1.00 %	0.98 %	0.99 %	
	August	1.26 %	1.07 %	1.06 %	1.08 %	1.08 %	1.09 %	1.07 %	1.07 %	
	September	1.89 %	1.01 %	0.9 %	0.94 %	0.96 %	0.95 %	0.95 %	0.92 %	
	Month	Order								
		First	Second	Third	Fourth	Fifth	Sixth	Seventh	Eighth	
Lee et al. (2007)	June	1.24 %	0.74 %	0.64 %	0.72 %	0.65 %	0.66 %	0.64 %	0.65 %	
	July	1.23 %	0.78 %	0.73 %	0.83 %	0.70 %	0.71 %	0.68 %	0.69 %	
	August	1.09 %	0.92 %	0.88 %	1.07 %	0.75 %	0.76 %	0.75 %	0.73 %	
	September	1.28 %	0.91 %	0.86 %	1.03 %	0.87 %	0.97 %	0.84 %	0.82 %	

(continued)

Table 5.12 (continued)

Model	Month	Order								
		First	Second	Third	Fourth	Fifth	Sixth	Seventh	Eighth	
	Annealing constant α	Month	Order							
			First	Second	Third	Fourth	Fifth	Sixth	Seventh	Eighth
Lee et al. (2008)	0.9	June	0.79 %	0.46 %	0.42 %	0.44 %	0.42 %	0.41 %	0.46 %	0.39 %
		July	0.62 %	0.46 %	0.45 %	0.44 %	0.44 %	0.41 %	0.40 %	0.40 %
		August	0.66 %	0.40 %	0.40 %	0.40 %	0.36 %	0.41 %	0.39 %	0.44 %
		September	0.62 %	0.59 %	0.61 %	0.57 %	0.54 %	0.59 %	0.57 %	0.50 %
	Month	Window Bases								
		w = 2	w = 3	w = 4	w = 5	w = 6	w = 7	w = 8		
Chang and Chen (2009)	June	1.70 %	1.50 %	1.38 %	1.37 %	1.28 %	1.13 %	0.97 %		
	July	1.62 %	1.77 %	1.74 %	1.68 %	1.77 %	1.72 %	3.04 %		
	August	1.60 %	1.48 %	1.24 %	1.30 %	1.28 %	2.41 %	2.97 %		
	September	1.44 %	1.51 %	1.35 %	1.20 %	2.02 %	2.49 %	2.42 %		
	Month	Order								
		First	Second	Third	Fourth	Fifth	Sixth	Seventh	Eighth	
Wang and Chen (2009)	June	0.53 %	0.28 %	0.29 %	0.30 %	0.29 %	0.29 %	0.28 %	0.29 %	
	July	0.71 %	0.34 %	0.35 %	0.34 %	0.34 %	0.35 %	0.33 %	0.32 %	
	August	0.32 %	0.23 %	0.22 %	0.22 %	0.22 %	0.23 %	0.23 %	0.22 %	
	September	0.74 %	0.51 %	0.49 %	0.51 %	0.51 %	0.53 %	0.5 %	0.51 %	

(shown in Table 5.11) for the main-factor and second-factor that exhibit the best behavior in terms of AFER.

The experimental results of our proposed model and existing competing models such as Chen and Hwang (2000), Lee (2006), Lee et al. (2007, 2008), Chang and Chen (2009) and Wang and Chen (2009) are presented in Table 5.12 in terms of AFERs. The comparative analyzes in Table 5.12 signify that our proposed model exhibits higher accuracy than those of considered competing models.[1]

5.5 Discussion

In this chapter, a new model is proposed for handling two-factors forecasting problems based on the hybridization of ANN with the FTS. For generation of intervals of time series data sets, the SOFM neural network is used. Then, some proposed rules of interval weighing are used to compute the forecasted values. From empirical analyzes of experimental results, it is evident that our model is superior compared to the considered competing models in terms of accuracy.

References

Chang YC, Chen SM (2009) Temperature prediction based on fuzzy clustering and fuzzy rules interpolation techniques. Proceedings of the 2009 IEEE international conference on systems, man, and cybernetics. San Antonio, TX, USA, pp 3444–3449

Chen SM, Hwang JR (2000) Temperature prediction using fuzzy time series. IEEE Trans Syst Man Cybern Part B Cybern 30:263–275

Kohonen T (1990) The self organizing maps. In: Proceedings of IEEE, vol 78, pp 1464–1480

Lee LW, Wang LH, Chen SM (2007) Temperature prediction and TAIFEX forecasting based on fuzzy logical relationships and genetic algorithms. Expert Syst Appl 33(3):539–550

Lee LW, Wang LH, Chen SM (2008) Temperature prediction and TAIFEX forecasting based on high-order fuzzy logical relationships and genetic simulated annealing techniques. Expert Syst Appl 34(1):328–336

Lee ZY (2006) Method of bilaterally bounded to solution blasius equation using particle swarm optimization. Appl Math Comput 179(2):779–786

Liao TW (2005) Clustering of time series data—a survey. Pattern Recogn 38(11):1857–1874

Sivanandam SN, Deepa SN (2007) Principles of soft computing. Wiley India (P) Ltd., New Delhi

Wang NY, Chen SM (2009) Temperature prediction and TAIFEX forecasting based on automatic clustering techniques and two-factors high-order fuzzy time series. Expert Syst Appl 36(2, Part 1):2143–2154

[1]References are: (Chen and Hwang 2000; Lee 2006; Lee et al. 2007, 2008; Chang and Chen 2009; Wang and Chen 2009).

Chapter 6
FTS-PSO Based Model for M-Factors Time Series Forecasting

It is not the answer that enlightens, but the question.
By E. I. Decouvertes (1909–1994)

Abstract In real-time, one observation always relies on several observations. To improve the forecasting accuracy, all these observations can be incorporated in forecasting models. Therefore, in this chapter, we have intended to introduce a new Type-2 FTS model that can utilize more observations in forecasting. Later, this Type-2 model is enhanced by employing PSO technique. The main motive behind the utilization of the PSO with the Type-2 model is to adjust the lengths of intervals in the universe of discourse that are employed in forecasting, without increasing the number of intervals. The daily stock index price data set of SBI (State Bank of India) is used to evaluate the performance of the proposed model. The proposed model is also validated by forecasting the daily stock index price of Google. Our experimental results demonstrate the effectiveness and robustness of the proposed model in comparison with existing FTS models and conventional time series models.

Keywords FTS · Stock index forecasting · Type-1 · Type-2 · PSO · Defuzzification

6.1 Background and Related Literature

The application of FTS in financial forecasting has also attracted researchers' attention in the recent years. Many researchers focus on designing the models for TAIEX[1] and TIFEX[2] forecasting. Their applications are limited to deal with either one-factor

---

[1]References are: (Liu and Wei 2010; Chang et al. 2011; Cheng et al. 2013; Wei 2013).
[2]References are: (Park et al. 2010; Hsu et al. 2010; Kuo et al. 2010; Aladag 2013).

P. Singh, *Applications of Soft Computing in Time Series Forecasting*,
Studies in Fuzziness and Soft Computing 330, DOI 10.1007/978-3-319-26293-2_6

or two-factors time series data sets. For the stock index forecasting, Huarng and Yu (2005) show that the forecasting accuracy can be improved by including more observations (e.g., *close, high*, and *low*) in the models.

All the models proposed by researchers above are based on Type-1 fuzzy set concept except the model proposed by Huarng and Yu (2005) in 2005, which is based on Type-2 fuzzy set concept. Researchers employ Type-2 fuzzy set concept (which is an extension of Type-1 fuzzy set) in various domains such as control system design and modeling (Pedrycz and Song 2012; Molaeezadeh and Moradi 2013), because Type-2 fuzzy set systems are much more powerful than Type-1 fuzzy sets systems to represent highly nonlinear and/or uncertain systems (Kumbasar et al. 2013). Nowadays, Type-2 fuzzy set concept is successfully applied in time series forecasting (Karnik et al. 1999; Mencattini et al. 2006; Almaraashi and John 2010, 2011).

In this study, we aim to propose an improved FTS model by employing M-factors time series data set. To deal with these factors together, we design a model based on Type-2 FTS concept, which is an improvement over the existing Type-2 model proposed by Huarng and Yu (2005). Later, to enhance the forecasting accuracy, we hybridize the PSO algorithm with the proposed Type-2 model. The daily stock index price data set of SBI is employed for the experimental purpose, which consists of 4-factors, *viz.*, "Open", "High", "Low" and "Close" factors/variables. After that, performance of the hybrid model is evaluated, which demonstrates its effectiveness over conventional FTS models and non-FTS models. The proposed model is also validated by forecasting the stock index price of Google.

6.2 Fuzzy Operators and Its Application

Following Definition 2.2.12, we define two operators, *viz.*, union and intersection, for the set theoretic operations. These two operators are explained as follows (Ross 2007).

Observation that is handled by Type-1 FTS model can be termed as "main-factor/Type-1 observation", whereas observations that are handled by Type-2 FTS model can be termed as "secondary-factors / Type-2 observations". Due to involvement of Type-2 observations with Type-1 observation, many FLRGs are generated in Type-2 model. To deal with these FLRGs together and establish the relationships among multiple FLRGs, the union ($\cup$) and intersection ($\cap$) operations are employed. Both these operations in terms of handling FLRGs are explained below.

Consider the following FLRGs for Type-1 and Type-2 observations as follows:

FLRG for Type-1 observation : $A_p \rightarrow A_{s1}, A_{s2}, \ldots, A_{sm}$.

$$\text{FLRGs for Type-2 observations} \begin{cases} A_q \rightarrow A_{t1}, A_{t2}, \ldots, A_{tm}, \\ A_r \rightarrow A_{u1}, A_{u2}, \ldots, A_{um}, \\ \qquad \ldots \end{cases}$$

Algorithm 4 Huarng and Yu Model for Type-2 FTS

Step 1: Choose a Type-1 FTS model.
Step 2: Pick a variable and Type-1 observations.
Step 3: Apply the Type-1 model to the Type-1 observations and obtain FLRGs.
Step 4: Pick Type-2 observations.
Step 5: Map out-of-sample observations to FLRGs and obtain forecasts.
Step 6: Apply operators to the FLRGs for all the observations.
Step 7: Defuzzify the forecasts.
Step 8: Calculate forecasts for Type-2 model.
Step 9: Evaluate the performance.

Based on these FLRGs, we define $\cup$ and $\cap$ operations as follows:

Definition 6.2.1 ($\cup$ *operation for FLRGs*). The operator $\cup$ is used to establish the relationships between the FLRGs of Type-1 and Type-2 observations by selecting the maximum value from the right-hand side of each of the FLRG as follows:

$$\begin{aligned} A_p, A_q, A_r, \ldots \rightarrow\ & \cup(A_{s1}, A_{s2}, \ldots A_{sm}), \\ & \cup(A_{t1}, A_{t2}, \ldots, A_{tm}), \\ & \cup(A_{u1}, A_{u2}, \ldots, A_{um}), \\ & \ldots \end{aligned} \tag{6.2.1}$$

Following Definition 6.2.1, the FLRGs of Type-1 and Type-2 observations are combined as follows:

$$\begin{aligned} A_p, A_q, A_r, \ldots \rightarrow\ & (A_{s1} \vee A_{s2} \vee \ldots \vee A_{sm}), \\ & (A_{t1} \vee A_{t2} \vee \ldots \vee A_{tm}), \\ & (A_{u1} \vee A_{u2} \vee \ldots \vee A_{um}), \\ & \ldots \end{aligned} \tag{6.2.2}$$

Definition 6.2.2 ($\cap$ *operation for FLRGs*). The operator $\cap$ is used to establish the relationships between the FLRGs of Type-1 and Type-2 observations by selecting the minimum value from the right-hand side of each of the FLRG as follows:

$$\begin{aligned} A_p, A_q, A_r, \ldots \rightarrow\ & \cap(A_{s1}, A_{s2}, \ldots A_{sm}), \\ & \cap(A_{t1}, A_{t2}, \ldots, A_{tm}), \\ & \cap(A_{u1}, A_{u2}, \ldots, A_{um}), \\ & \ldots \end{aligned} \tag{6.2.3}$$

Following Definition 6.2.2, the FLRGs of Type-1 and Type-2 observations are combined as follows:

$$\begin{aligned} A_p, A_q, A_r, \ldots \rightarrow\ & (A_{s1} \wedge A_{s2} \wedge \ldots \wedge A_{sm}), \\ & (A_{t1} \wedge A_{t2} \wedge \ldots \wedge A_{tm}), \\ & (A_{u1} \wedge A_{u2} \wedge \ldots \wedge A_{um}), \\ & \ldots \end{aligned} \tag{6.2.4}$$

6.3 Algorithm and Defuzzification for Type-2 Model

In this section, we will first present the algorithm for the proposed Type-2 FTS model. Then, defuzzification process for the proposed model will be presented in the subsequent subsection.

6.3.1 Algorithm

In this subsection, we have presented an algorithm for the Type-2 FTS model, which is based on Huarng and Yu (2005) model. Therefore, we first present the algorithm for Type-2 model proposed by Huarng and Yu (2005). This algorithm is presented as Algorithm 4. To improve the forecasting accuracy of Type-2 FTS model, we apply some changes in the Algorithm 4. This algorithm is presented as Algorithm 5.

Algorithm 5 Proposed Type-2 FTS Forecasting Model

Step 1: Select Type-1 and Type-2 observations.
Step 2: Determine the universe of discourse of time series data set and partition it into equal lengths of intervals.
Step 3: Define linguistic terms for each of the interval.
Step 4: Fuzzify the time series data set of Type-1 and Type-2 observations.
Step 5: Establish the FLRs based on Definition 2.2.7.
Step 6: Construct the FLRGs based on Definition 2.2.8.
Step 7: Establish the relationships between FLRGs of both Type-1 and Type-2 observations, and map-out them to their corresponding day.
Step 8: Based on Definitions 6.2.1 and 6.2.2, apply $\cup$ and $\cap$ operators on mapped-out FLRGs of Type-1 and Type-2 observations, and obtain the fuzzified forecasting data.
Step 9: Defuzzify the forecasting data based on the "Frequency-Weighing Defuzzification Technique".
Step 10: Compute the forecasted values individually for $\cup$ and $\cap$ operations.

Table 6.1 A sample of daily stock index price list of SBI (in rupee)

Date (mm/dd/yy)	Open	High	Low	Close
6/4/2012	2064.00	2110.00	2061.00	2082.75
6/5/2012	2100.00	2168.00	2096.00	2158.25
6/6/2012	2183.00	2189.00	2156.55	2167.95
6/7/2012	2151.55	2191.60	2131.50	2179.45
6/10/2012	2200.00	2217.90	2155.50	2164.80
6/11/2012	2155.00	2211.85	2132.60	2206.15
...	...	...	...	...
7/29/2012	1951.25	2040.00	1941.00	2031.75

6.3.2 *Defuzzification for Type-2 Model*

After obtaining the fuzzified forecasting data from the $\cup$ and $\cap$ operations, they are defuzzified based on the "Frequency-Weighing Defuzzification Technique (FWDT)", which is the modified version of the defuzzification technique proposed by Singh and Borah (2012). In the subsequent section, this technique is discussed. The defuzzified values obtained for the left-hand side and right-hand side of the FLRGs can be referred to as points "M" and "N" in Fig. 2.1 (Chap. 2), respectively. The forecasted value of the proposed Type-2 model can be derived by taking the average of these defuzzified values. The forecasted value for this Type-2 model can be referred to as point "D" in Fig. 2.1 (Chap. 2).

6.4 Type-2 FTS Forecasting Model

In this section, the proposed "Type-2 FTS Forecasting Model" is presented. To verify the proposed model, the daily stock index price data set of SBI for the period 6/4/2012–7/29/2012 (format: mm/dd/yy), is collected from the website.[3] A sample of data set is listed in Table 6.1. The model consists of ten phases. The functionality of each phase is explained below.

Step 6.4.1 Select Type-1 and Type-2 observations.

[Explanation] In this study, we select "Actual Price" as the main forecasting objective. To obtain the forecasting results, the "Close" variable of the stock index data set as shown in Table 6.1, has been selected as Type-1 observation, whereas "Open", "High" and "Low" variables have been selected as Type-2 observations.

Step 6.4.2 Determine the universe of discourse of time series data set and partition it into equal lengths of intervals.

---

[3]http://in.finance.yahoo.com.

[Explanation] Define the universe of discourse U of the historical time series data set. Assume that $U = [M_{min} - F_1, M_{max} + F_2]$, where M_{min} and M_{max} are the minimum and maximum values of the historical time series data set. Here, F_1 and F_2 are two positive numbers. These M_{min} and M_{max} values are obtained by considering both Type-1 (Closing price) and Type-2 (Open, High and Low) observations. Based on Table 6.1, we can see that $M_{min} = 1931.50$ and $M_{max} = 2252.55$. Therefore, we let $F_1 = 2$ and $F_2 = 0.50$. Thus, in this study, the universe of discourse $U = [1929.50, 2253.10]$.

Now, divide the universe of discourse U into n equal lengths of intervals (i.e., $a_1, a_2, \ldots,$ and a_n). Each interval of time series data set can be defined as follows (Lee and Chou 2004):

$$a_i = [L_B + (i-1)\frac{U_B - L_B}{j}, L_B + i\frac{U_B - L_B}{j}] \tag{6.4.1}$$

for $i = 1, 2, \ldots, n$ and $j = 30$. Here, $L_B = 1929.50$ and $U_B = 2253.10$. Here, j represents the number of intervals which are taken into the consideration.

Based on the Eq. 6.4.1, the universe of discourse $U = [1929.50, 2253.10]$ is divided into 30 equal lengths of intervals as: $a_1 = (1929.50, 1940.30]$, $a_2 = (1940.30, 1951.10]$, ..., $a_{30} = (2242.30, 2253.10]$. Now, assign the data to their corresponding intervals. Mid-value of each interval is recorded by taking the mean of lower bound and upper bound of the interval. For ease of computation, intervals which do not cover any historical datum is discarded from the list. Intervals which contain historical data are represented as I_i, for $i = 1, 2, \ldots, n$; and $n \leq 26$. Then, each interval is assigned a weight based on frequency of the interval. For example, in Table 6.2, interval I_2 has two data with frequency 2. So, we assign weight 2 to the interval I_2. All these intervals, and their corresponding data, mid-values and weights are shown in Table 6.2.

Step 6.4.3 Define linguistic terms for each of the interval. Assume that the historical time series data set is distributed among n intervals (i.e., $I_1, I_2, \ldots,$ and I_n). Therefore, define n linguistic variables $A_1, A_2, \ldots, A_n$, which can be represented by fuzzy sets, as shown below:

$$\begin{aligned}
A_1 &= 1/I_1 + 0.5/I_2 + 0/I_3 + \ldots + 0/I_{n-2} + 0/I_{n-1} + 0/I_n, \\
A_2 &= 0.5/I_1 + 1/I_2 + 0.5/I_3 + \ldots + 0/I_{n-2} + 0/I_{n-1} + 0/I_n, \\
A_3 &= 0/I_1 + 0.5/I_2 + 1/I_3 + \ldots + 0/I_{n-2} + 0/I_{n-1} + 0/I_n, \\
&\vdots \\
A_n &= 0/I_n + 0/I_2 + 0/I_3 + \ldots + 0/I_{n-2} + 0.5/I_{n-1} + 1/I_n.
\end{aligned}$$

Here, maximum degree of membership of the fuzzy set A_i occurs at interval I_i, and $1 \leq i \leq n$.

Table 6.2 A sample of intervals, and their corresponding data, mid-points and weights

Interval	Corresponding data	Mid-value	Weight
$I_1 = (1929.50, 1940.30]$	Open={Nil}	1934.90	1
	High={Nil}		
	Low={1931.50}		
	Close={Nil}		
$I_2 = (1940.30, 1951.10]$	Open={Nil}	1945.7.60	2
	High={Nil}		
	Low={1941.00}		
	Close={1941.00}		
$I_3 = (1951.10, 1961.90]$	Open={1951.25}	1956.50	1
	High={Nil}		
	Low={Nil}		
	Close={Nil}		
...	...	...	...
$I_{26} = (2242.30, 2253.10]$	Open={Nil}	2247.70	2
	High={2244.00, 2252.55}		
	Low={Nil}		
	Close={Nil}		

[Explanation] We define 26 linguistic variables $A_1, A_2, \ldots, A_{26}$ for the stock index data set, because the data set is distributed among 26 intervals (i.e., $I_1, I_2, \ldots,$ and I_{26}). All these defined linguistic variables are shown as below:

$$\begin{aligned}
A_1 &= 1/I_1 + 0.5/I_2 + 0/I_3 + \ldots + 0/I_{n-2} + 0/I_{n-1} + 0/I_{26}, \\
A_2 &= 0.5/I_1 + 1/I_2 + 0.5/I_3 + \ldots + 0/I_{n-2} + 0/I_{n-1} + 0/I_{26}, \\
A_3 &= 0/I_1 + 0.5/I_2 + 1/I_3 + \ldots + 0/I_{n-2} + 0/I_{n-1} + 0/I_{26}, \\
&\vdots \\
A_{26} &= 0/I_1 + 0/I_2 + 0/I_3 + \ldots + 0/I_{n-2} + 0.5/I_{n-1} + 1/I_{26}. \qquad (6.4.2)
\end{aligned}$$

For ease of computation, the degree of membership values of fuzzy set $A_j (j = 1, 2, \ldots, 26)$ are considered as either 0, 0.5 or 1, and $1 \leq j \leq 26$. In Eq. 6.4.2, for example, A_1 represents a linguistic value, which denotes a fuzzy set on $\{I_1, I_2, \ldots, I_{26}\}$. This fuzzy set consists of twenty six members with different degree of membership values $= \{1, 0.5, 0, \ldots, 0\}$. Similarly, the linguistic value A_2 denotes the fuzzy set on $\{I_1, I_2, \ldots, I_{26}\}$, which also consists of twenty six members with different degree of membership values $= \{0.5, 1, 0.5, \ldots, 0\}$. The descriptions of remaining linguistic variables, *viz.*, $A_3, A_4, \ldots, A_{26}$, can be provided in a similar manner.

Step 6.4.4 Fuzzify the time series data set. If one day's datum belongs to the interval I_i, then datum is fuzzified into A_i, where $1 \leq i \leq n$.

[Explanation] In Step 6.4.3, each fuzzy set contains twenty six intervals, and each interval corresponds to all fuzzy sets with different degree of membership values. For example, interval I_1 corresponds to linguistic variables A_1 and A_2 with degree of membership values 1 and 0.5, respectively, and remaining fuzzy sets with degree of membership value 0. Similarly, interval I_2 corresponds to linguistic variables A_1, A_2 and A_3 with degree of membership values 0.5, 1, and 0.5, respectively, and remaining fuzzy sets with degree of membership value 0. The descriptions of remaining intervals, *viz.*, $I_3, I_4, \ldots, I_{26}$, can be provided in a similar manner.

In order to fuzzify the historical time series data, it is essential to obtain the degree of membership value of each observation belonging to each A_j $(j = 1, 2, \ldots, n)$ for each day. If the maximum membership value of one day's observation occurs at interval I_i and $1 \leq i \leq n$, then the fuzzified value for that particular day is considered as A_i. For example, the stock index price of 6/4/2012 for observation "Open" belongs to the interval I_9 with the highest degree of membership value 1, so it is fuzzified to A_9. In this way, we have fuzzified each observation of the historical time series data set. The fuzzified historical time series data set is presented in Table 6.3.

Step 6.4.5 Establish the FLR between the fuzzified values obtained in Step 6.4.4. For example, if the fuzzified values of time $t-1$ and t are A_i and A_j, respectively, then establish the FLR as "$A_i \rightarrow A_j$", where "A_i" and "A_j" are called the previous state and the current state of the FLR, respectively.

[Explanation] In Table 6.3, fuzzified stock index values for "Open" observation for days 6/4/2012 and 6/5/2012 are A_9 and A_{12}, respectively. So, we can establish a FLR between two consecutive fuzzified values A_9 and A_{12} as "$A_9 \rightarrow A_{12}$", where "A_9" and "A_{12}" are called the previous state and current state of the FLR, respectively. In this way, we have obtained all FLRs for both Type-1 and Type-2 fuzzified stock index values, which are presented in Table 6.4.

Table 6.3 A sample of fuzzified stock index data set

Date (mm/dd/yy)	Open	Fuzzified open	High	Fuzzified high	Low	Fuzzified low	Close	Fuzzified close
6/4/2012	2064.00	A_9	2110.00	A_{13}	2061.00	A_9	2082.80	A_{11}
6/5/2012	2100.00	A_{12}	2168.00	A_{19}	2096.00	A_{12}	2158.30	A_{18}
6/6/2012	2183.00	A_{20}	2189.00	A_{21}	2156.60	A_{18}	2167.90	A_{19}
6/7/2012	2151.60	A_{17}	2191.60	A_{21}	2131.50	A_{15}	2179.40	A_{20}
6/10/2012	2200.00	A_{22}	2217.90	A_{23}	2155.50	A_{17}	2164.80	A_{18}
6/11/2012	2155.00	A_{17}	2211.80	A_{23}	2132.60	A_{15}	2206.20	A_{22}
...	...	...	...	...	...	...	...	...
7/29/2012	1951.30	A_3	2040.00	A_7	1941.00	A_2	2031.80	A_6

Table 6.4 A sample of FLRs

FLRs for Open	FLRs for High	FLRs for Low	FLRs for Close
$A_9 \to A_{12}$	$A_{13} \to A_{19}$	$A_9 \to A_{12}$	$A_{11} \to A_{18}$
$A_{12} \to A_{20}$	$A_{19} \to A_{21}$	$A_{12} \to A_{18}$	$A_{18} \to A_{19}$
...	...	...	...
$A_{12} \to A_{17}$	$A_{20} \to A_{20}$	$A_{11} \to A_{15}$	$A_{20} \to A_{18}$
$A_{12} \to A_{11}$	$A_{14} \to A_{12}$	$A_{10} \to A_9$	$A_{12} \to A_{10}$
...	...	...	...

Table 6.5 A sample of FLRGs

FLRGs for Open	FLRGs for High	FLRGs for Low	FLRGs for Close
$A_8 \to A_3$	$A_9 \to A_7$	$A_1 \to A_2$	$A_2 \to A_6$
$A_9 \to A_{12}$	$A_{11} \to A_9$	$A_4 \to A_1$	$A_5 \to A_2$
$A_{10} \to A_8$	$A_{12} \to A_{11}$	$A_8 \to A_{12}$	$A_{10} \to A_5$
...	...	...	...
$A_{25} \to A_{21}$		$A_{23} \to A_{21}, A_{22}, A_{23}$	

Step 6.4.6 Construct the FLRG. Based on the same previous state of the FLRs, the FLRs can be grouped into an FLRG. For example, the FLRG "$A_i \to A_m, A_n$" indicates that there are the following FLRs:

$$A_i \to A_m,$$
$$A_i \to A_n.$$

[Explanation] In Table 6.4, there are 3 FLRs for "Open" observation with the same previous state, $A_{12} \to A_{11}$, $A_{12} \to A_{17}$, and $A_{12} \to A_{20}$. These FLRs are used to form the FLRG, $A_{12} \to A_{11}, A_{17}, A_{20}$. All these FLRGs are shown in Table 6.5. If the same FLR appears more than once, it is included only once in the group.

Step 6.4.7 Establish the relationships between Type-1 and Type-2 observations. If one day's fuzzified stock index price value for Type-1 observation is A_i with FLRG "$A_i \to A_j$", then FLRG is mapped-out to its corresponding day. For the same day, if fuzzified stock index price values for the three different Type-2 observations are A_k, A_m and A_n with FLRGs "$A_k \to A_a$, $A_m \to A_b$, and $A_n \to A_c$" respectively, then all these FLRGs are also mapped-out to that day.

[Explanation] In Table 6.5, the first three columns represent the FLRGs of Type-2 observations, and the last column represents the FLRGs of Type-1 observation. To establish the relationship between Type-1 and Type-2 observations, FLRGs of both Type-1 and Type-2 observations are mapped-out to their corresponding day. For example, fuzzified stock index price value for Type-1 observation "Close" for day 6/5/2012 is A_{18}. FLRG of this fuzzy set is:
FLRG for Type-1 observation : $A_{18} \to A_{14}, A_{15}, A_{19}, A_{20}, A_{22}$.

Similarly, fuzzified stock index price values for Type-2 observations "Open", "High" and "Low" for day 6/5/2012 are A_{12}, A_{19} and A_{12}, respectively. FLRGs for these fuzzy sets are shown below:

$$\text{FLRG for Type-2 observations} \begin{cases} A_{12} \rightarrow A_{11}, A_{17}, A_{20} \\ A_{19} \rightarrow A_{15}, A_{21} \\ A_{12} \rightarrow A_{11}, A_{12}, A_{13}, A_{18} \end{cases}$$

FLRGs of both Type-1 and Type-2 observations are now mapped-out together to day 6/5/2012. In this way, we have mapped-out all FLRGs to their corresponding day, which are shown in Table 6.6.

Step 6.4.8 Obtain the fuzzified forecasting data by applying $\cup$ and $\cap$ operators on mapped-out FLRGs of both Type-1 and Type-2 observations.

[Explanation] Based on Definitions 6.2.1 and 6.2.2, obtain the forecasting data by applying $\cup$ and $\cap$ operators on mapped-out FLRGs of both Type-1 and Type-2 observations. The mapped-out FLRGs are shown in Table 6.6. In Table 6.6, we apply $\cup$ operator to all mapped-out FLRGs of both Type-1 and Type-2 observations. For example, mapped-out FLRG of Type-1 observation for day 6/7/2012 is:

$$A_{20} \rightarrow A_{11}, A_{18}, A_{20}, A_{22}, A_{24} \text{ (for Close)}$$

and, mapped-out FLRGs of Type-2 observations for day 6/7/2012 are:

$$A_{17} \rightarrow A_{13}, A_{18}, A_{22}, A_{23} \text{ (for Open)}$$
$$A_{21} \rightarrow A_{21}, A_{22}, A_{23}, A_{25} \text{ (for High)}$$
$$A_{15} \rightarrow A_{11}, A_{12}, A_{17}, A_{18}, A_{20} \text{ (for Low)}$$

Hence, from Definition 6.2.1, we have:

$$\begin{aligned} A_{20}, A_{17}, A_{21}, A_{15} \rightarrow\ & \cup(A_{11}, A_{18}, A_{20}, A_{22}, A_{24}), \\ & \cup(A_{13}, A_{18}, A_{22}, A_{23}), \\ & \cup(A_{21}, A_{22}, A_{23}, A_{25}), \\ & \cup(A_{11}, A_{12}, A_{17}, A_{18}, A_{20}) \end{aligned}$$

Now, based on Eq. 6.2.2, above mapped-out FLRGs can be represented in the following form as follows:

$$\begin{aligned} A_{20}, A_{17}, A_{21}, A_{15} \rightarrow\ & (A_{11} \vee A_{18} \vee A_{20} \vee A_{22} \vee A_{24}), \\ & (A_{13} \vee A_{18} \vee A_{22} \vee A_{23}), \\ & (A_{21} \vee A_{22} \vee A_{23} \vee A_{25}), \\ & (A_{11} \vee A_{12} \vee A_{17} \vee A_{18} \vee A_{20}) \end{aligned}$$

i.e., $A_{20}, A_{17}, A_{21}, A_{15} \rightarrow A_{20}, A_{23}, A_{24}, A_{25}$.

Table 6.6 A sample of mapped-out FLRGs with their corresponding day

Date (mm/dd/yy)	FLRGs of Type-2 observation (open)	FLRGs of Type-2 observation (high)	FLRGs of Type-2 observation (low)	FLRGs of Type-1 observation (close)
6/4/2012	$A_9 \rightarrow A_{12}$	$A_{13} \rightarrow A_{19}$	$A_9 \rightarrow A_4, A_{12}$	$A_{11} \rightarrow A_{12}, A_{18}$
6/5/2012	$A_{12} \rightarrow A_{11}, A_{17}, A_{20}$	$A_{19} \rightarrow A_{15}, A_{21}$	$A_{12} \rightarrow A_{11}, A_{12}, A_{13}, A_{18}$	$A_{18} \rightarrow A_{14}, A_{15}, A_{19}, A_{20}, A_{22}$
6/6/2012	$A_{20} \rightarrow A_{17}, A_{23}$	$A_{21} \rightarrow A_{21}, A_{22}, A_{23}, A_{25}$	$A_{18} \rightarrow A_{15}, A_{17}, A_{19}$	$A_{19} \rightarrow A_{20}$
6/7/2012	$A_{17} \rightarrow A_{13}, A_{18}, A_{22}, A_{23}$	$A_{21} \rightarrow A_{21}, A_{22}, A_{23}, A_{25}$	$A_{15} \rightarrow A_{11}, A_{12}, A_{17}, A_{18}, A_{20}$	$A_{20} \rightarrow A_{11}, A_{18}, A_{20}, A_{22}, A_{24}$
6/10/2012	$A_{22} \rightarrow A_{17}$	$A_{23} \rightarrow A_{19}, A_{21}, A_{23}, A_{24}, A_{25}, A_{26}$	$A_{17} \rightarrow A_{11}, A_{15}$	$A_{18} \rightarrow A_{14}, A_{15}, A_{19}, A_{20}, A_{22}$
6/11/2012	$A_{17} \rightarrow A_{13}, A_{18}, A_{22}, A_{23}$	$A_{23} \rightarrow A_{19}, A_{21}, A_{23}, A_{24}, A_{25}, A_{26}$	$A_{15} \rightarrow A_{11}, A_{12}, A_{17}, A_{18}, A_{20}$	$A_{22} \rightarrow A_{21}, A_{24}$
6/12/2012	$A_{23} \rightarrow A_{11}, A_{17}, A_{19}, A_{21}, A_{23}, A_{25}$	$A_{26} \rightarrow A_{23}, A_{25}$	$A_{20} \rightarrow A_{16}, A_{18}, A_{20}$	$A_{24} \rightarrow A_{17}, A_{20}, A_{22}, A_{23}, A_{25}$
6/13/2012	$A_{23} \rightarrow A_{11}, A_{17}, A_{19}, A_{21}, A_{23}, A_{25}$	$A_{23} \rightarrow A_{19}, A_{21}, A_{23}, A_{24}, A_{25}, A_{26}$	$A_{16} \rightarrow A_{17}$	$A_{17} \rightarrow A_{20}$
6/14/2012	$A_{19} \rightarrow A_{21}, A_{23}$	$A_{21} \rightarrow A_{21}, A_{22}, A_{23}, A_{25}$	$A_{17} \rightarrow A_{11}, A_{15}$	$A_{20} \rightarrow A_{11}, A_{18}, A_{20}, A_{22}, A_{24}$
6/17/2012	$A_{23} \rightarrow A_{11}, A_{17}, A_{19}, A_{21}, A_{23}, A_{25}$	$A_{25} \rightarrow A_{14}, A_{23}, A_{25}, A_{26}$	$A_{11} \rightarrow A_8, A_{10}, A_{15}$	$A_{11} \rightarrow A_{12}, A_{18}$
6/18/2012	$A_{11} \rightarrow A_{10}, A_{14}$	$A_{14} \rightarrow A_{12}, A_{15}$	$A_8 \rightarrow A_{12}$	$A_{12} \rightarrow A_{10}, A_{12}, A_{14}, A_{18}$
6/19/2012	$A_{14} \rightarrow A_{12}, A_{15}$	$A_{15} \rightarrow A_{14}, A_{16}, A_{19}, A_{20}$	$A_{12} \rightarrow A_{11}, A_{12}, A_{13}, A_{18}$	$A_{14} \rightarrow A_{12}, A_{14}, A_{20}$
6/20/2012	$A_{12} \rightarrow A_{11}, A_{17}, A_{20}$	$A_{20} \rightarrow A_{20}, A_{22}$	$A_{11} \rightarrow A_8, A_{10}, A_{15}$	$A_{20} \rightarrow A_{11}, A_{18}, A_{20}, A_{22}, A_{24}$
6/21/2012	$A_{17} \rightarrow A_{13}, A_{18}, A_{22}, A_{23}$	$A_{20} \rightarrow A_{20}, A_{22}$	$A_{15} \rightarrow A_{11}, A_{12}, A_{17}, A_{18}, A_{20}$	$A_{18} \rightarrow A_{14}, A_{15}, A_{19}, A_{20}, A_{22}$
6/24/2012	$A_{18} \rightarrow A_{14}$	$A_{22} \rightarrow A_{15}, A_{23}, A_{24}$	$A_{12} \rightarrow A_{11}, A_{12}, A_{13}, A_{18}$	$A_{14} \rightarrow A_{12}, A_{14}, A_{20}$
...	...	...	...	...
7/29/2012	$A_3 \rightarrow ?$	$A_7 \rightarrow ?$	$A_2 \rightarrow ?$	$A_6 \rightarrow ?$

Similarly, in Table 6.6, we apply $\cap$ operator to mapped-out FLRGs of Type-1 and Type-2 observations for day 6/7/2012. Hence, from Definition 6.2.2, we have:

$$\begin{aligned} A_{20}, A_{17}, A_{21}, A_{15} \rightarrow\ & \cap(A_{11}, A_{18}, A_{20}, A_{22}, A_{24}), \\ & \cap(A_{13}, A_{18}, A_{22}, A_{23}), \\ & \cap(A_{21}, A_{22}, A_{23}, A_{25}), \\ & \cap(A_{11}, A_{12}, A_{17}, A_{18}, A_{20}) \end{aligned}$$

Now, based on Eq. 6.2.4, above mapped-out FLRGs can be represented in the following form as follows:

$$\begin{aligned} A_{20}, A_{17}, A_{21}, A_{15} \rightarrow\ & (A_{11} \wedge A_{18} \wedge A_{20} \wedge A_{22} \wedge A_{24}), \\ & (A_{13} \wedge A_{18} \wedge A_{22} \wedge A_{23}), \\ & (A_{21} \wedge A_{22} \wedge A_{23} \wedge A_{25}), \\ & (A_{11} \wedge A_{12} \wedge A_{17} \wedge A_{18} \wedge A_{20}) \end{aligned}$$

i.e., $A_{15}, A_{17}, A_{20}, A_{21} \rightarrow A_{11}, A_{13}, A_{21}$.

Repeated fuzzy set is discarded from the mapped-out FLRGs. The fuzzified forecasting data obtained after applications of the $\cup$ and $\cap$ operators are presented in Tables 6.7 and 6.8, respectively.

Step 6.4.9 Defuzzify the forecasting data.

Table 6.7 A sample of fuzzified forecasting data for the $\cup$ operation

Date (mm/dd/yy)	Forecasting data – $\cup$ operation
6/4/2012	$A_9, A_{11}, A_{13} \rightarrow A_{12}, A_{18}, A_{19}$
6/5/2012	$A_{12}, A_{18}, A_{19} \rightarrow A_{18}, A_{20}, A_{21}, A_{22}$
6/6/2012	$A_{18}, A_{19}, A_{20}, A_{21} \rightarrow A_{19}, A_{20}, A_{23}, A_{25}$
6/7/2012	$A_{15}, A_{17}, A_{20}, A_{21} \rightarrow A_{20}, A_{23}, A_{24}, A_{25}$
...	...
7/29/2012	$A_2, A_3, A_6, A_7 \rightarrow ?$

Table 6.8 A sample of fuzzified forecasting data for the $\cap$ operation

Date (mm/dd/yy)	Forecasting data – $\cap$ operation
6/4/2012	$A_9, A_{11}, A_{13} \rightarrow A_4, A_{12}, A_{19}$
6/5/2012	$A_{12}, A_{18}, A_{19} \rightarrow A_{11}, A_{14}, A_{15}$
6/6/2012	$A_{18}, A_{19}, A_{20}, A_{21} \rightarrow A_{15}, A_{17}, A_{20}, A_{21}$
6/7/2012	$A_{15}, A_{17}, A_{20}, A_{21} \rightarrow A_{11}, A_{13}, A_{21}$
...	...
7/29/2012	$A_2, A_3, A_6, A_7 \rightarrow ?$

To defuzzify the fuzzified time series data set and to obtain the forecasted values, defuzzification technique proposed by Singh and Borah (Singh and Borah 2012) is employed here. Based on the application of technique, it is slightly modified and categorized as: **Principle 1** and **Principle 2**. The procedure for **Principle 1** is given as follows:

- **Principle 1**: The **Principle 1** is applicable only if there are more than one fuzzified values available in the current state. The steps under **Principle 1** are explained below.

Step 1 Obtain the fuzzified forecasting data for forecasting day $D(t)$, whose previous state is $A_{i1}, A_{i2}, \ldots, A_{ip}$ $(i = 1, 2, 3, \ldots, n)$, and current state is $A_{j1}, A_{j2}, \ldots, A_{jp}$ $(j = 1, 2, 3, \ldots, n)$, i.e., FLRG is in the form of $A_{i1}, A_{i2}, \ldots, A_{ip} \rightarrow A_{j1}, A_{j2}, \ldots, A_{jp}$.

Step 2 Obtain the defuzzified forecasting value for the previous state as:

$$\begin{aligned} Defuzz_{prev} = C_{i1} \times & \left[\frac{W_{i1}}{W_{i1} + W_{i2} + \ldots + W_{ip}} \right] + \\ C_{i2} \times & \left[\frac{W_{i2}}{W_{i1} + W_{i2} + \ldots + W_{ip}} \right] + \\ & \ldots \\ C_{ip} \times & \left[\frac{W_{ip}}{W_{i1} + W_{i2} + \ldots + W_{ip}} \right]. \end{aligned} \tag{6.4.3}$$

where C_{i1}, C_{i2}, …, C_{ip} and W_{i1}, W_{i2}, …, W_{ip} denote mid-values and weights of the intervals I_{i1}, I_{i2}, …, I_{ip} $(i = 1, 2, 3, \ldots, n)$, respectively, and the maximum membership values of A_{i1}, A_{i2}, …, A_{ip} occur at intervals I_{i1}, I_{i2}, …, I_{ip}, respectively.

Step 3 Obtain the defuzzified forecasting value for the current state as:

$$\begin{aligned} Defuzz_{curr} = C_{j1} \times & \left[\frac{W_{j1}}{W_{j1} + W_{j2} + \ldots + W_{jp}} \right] + \\ C_{j2} \times & \left[\frac{W_{j2}}{W_{j1} + W_{j2} + \ldots + W_{jp}} \right] + \\ & \ldots \\ C_{jp} \times & \left[\frac{W_{jp}}{W_{j1} + W_{j2} + \ldots + W_{jp}} \right]. \end{aligned} \tag{6.4.4}$$

where C_{j1}, C_{j2}, …, C_{jp} and W_{j1}, W_{j2}, …, W_{jp} denote mid-values and weights of the intervals I_{j1}, I_{j2}, …, I_{jp} $(j = 1, 2, 3, \ldots, n)$, respectively, and the maximum membership values of A_{j1}, A_{j2}, …, A_{jp} occur at intervals I_{j1}, I_{j2}, …, I_{jp}, respectively.

- **Principle 2**: This principle is applicable only if there is an unknown value in the current state. The steps under **Principle 2** are given as follows:

Step 1 Obtain the fuzzified forecasting data for forecasting day $D(t)$, whose previous state is $A_{i1}, A_{i2}, \ldots, A_{ip}$ $(i = 1, 2, 3, \ldots, n)$, and current state is "?" (the symbol "?" represents an unknown value), i.e., FLRG is in the form of $A_{i1}, A_{i2}, \ldots, A_{ip} \rightarrow ?$.

Step 2 Obtain the defuzzified forecasting value for the previous state as:

$$\begin{aligned} Defuzz_{prev} = C_{i1} \times & \left[\frac{W_{i1}}{W_{i1} + W_{i2} + \ldots + W_{ip}}\right] + \\ C_{i2} \times & \left[\frac{W_{i2}}{W_{i1} + W_{i2} + \ldots + W_{ip}}\right] + \\ & \cdots \\ C_{ip} \times & \left[\frac{W_{ip}}{W_{i1} + W_{i2} + \ldots + W_{ip}}\right]. \end{aligned} \tag{6.4.5}$$

where $C_{i1}, C_{i2}, \ldots, C_{ip}$ and $W_{i1}, W_{i2}, \ldots, W_{ip}$ denote mid-values and weights of the intervals $I_{i1}, I_{i2}, \ldots, I_{ip} (i = 1, 2, 3, \ldots, n)$, respectively.

Step 6.4.10 Compute the forecasted value for Type-2 FTS model.

If **Principle 1** is applicable, then forecasted value for day $D(t)$ can be computed as:

$$Forecast_{D(t)} = \frac{Defuzz_{prev} + Defuzz_{curr}}{2} \tag{6.4.6}$$

If **Principle 2** is applicable, then forecasted value for day $D(t)$ can be computed as:

$$Forecast_{D(t)} = Defuzz_{prev} \tag{6.4.7}$$

In this way, we obtain the forecasted values for the $\cup$ and $\cap$ operations individually based on the proposed model. To measure the performance of the model, AFER is used as an evaluation criterion. The AFER value of the forecasted stock index price is presented in Table 6.9.

Based on the proposed model, we present here an example to compute the forecasted value of daily stock index price of SBI using the $\cup$ operation as follows:

[Example] Suppose, we want to forecast the stock index price on day, $D(6/4/2012)$. To compute this value, obtain the fuzzified forecasting data for $D(6/4/2012)$ from Table 6.7, which is $A_9, A_{11}, A_{13} \rightarrow A_{12}, A_{18}, A_{19}$. Now, find the intervals where the maximum membership values for the previous state of FLRG (i.e., A_9, A_{11} and A_{13}) occur from Table 6.2, which are (I_9, I_{11} and I_{13}), respectively. The corresponding mid-values and weights for the intervals (I_9, I_{11} and I_{13}) are (2064.30, 2085.90 and 2107.50) and (4, 8 and 3), respectively.

Similarly, find the intervals where the maximum membership values for the current state of FLRG (i.e., A_{12}, A_{18}, and A_{19}) occur from Table 6.2, which are (I_{12}, I_{18} and I_{19}), respectively. The corresponding mid-values and weights for the intervals (I_{12}, I_{18} and I_{19}) are (2096.70, 2161.40 and 2172.20) and (12, 9 and 8), respectively.

Table 6.9 A sample of forecasted results of the stock index price of SBI (in rupee).

Date (mm/dd/yy)	Actual price (in rupee)	Proposed model ($\cup$ operation)	Proposed model ($\cap$ operation)
6/4/2012	2079.40	2111.10	2102.90
6/5/2012	2130.60	2161.60	2125.70
6/6/2012	2174.10	2191.00	2173.20
6/7/2012	2163.50	2189.00	2156.80
6/10/2012	2184.60	2176.40	2160.60
6/11/2012	2176.40	2193.90	2166.50
6/12/2012	2216.50	2205.30	2188.00
6/13/2012	2181.30	2191.90	2172.40
...	...	...	...
7/22/2012	2103.90	2124.60	2101.80
7/23/2012	2096.30	2129.90	2093.40
7/24/2012	2078.10	2091.80	2079.30
7/25/2012	2046.30	2047.00	2047.00
7/26/2012	1997.70	2002.80	2002.80
7/29/2012	1991.00	1969.40	1969.40
AFER		0.68 %	0.66 %

Now, obtain the defuzzified forecasting value for the previous state of the FLRG based on Eq. 6.4.3, which is equal to

$$\begin{aligned} Defuzz_{prev} &= 2064.30 \times \left[\frac{4}{4+8+3}\right] + 2085.90 \times \left[\frac{8}{4+8+3}\right] + \\ &\quad 2107.50 \times \left[\frac{3}{4+8+3}\right] \\ &= 2084.50 \end{aligned}$$

Similarly, obtain the defuzzified forecasting value for the current state of the FLRG based on Eq. 6.4.4, which is equal to

$$\begin{aligned} Defuzz_{curr} &= 2096.70 \times \left[\frac{12}{12+9+8}\right] + 2161.40 \times \left[\frac{9}{12+9+8}\right] + \\ &\quad 2172.20 \times \left[\frac{8}{12+9+8}\right] \\ &= 2137.60 \end{aligned}$$

Here, **Principle 1** is applicable to compute the forecasted value, because the current state of the FLRG does not contain any unknown value. Therefore, based on Eq. 6.4.6, the forecasted value for $D(6/4/2012)$ is equal to

$$Forecast_{D(6/4/2012)} = \left[\frac{2084.50 + 2137.60}{2}\right] = 2111.10$$

The forecasted results based on the proposed Type-2 FTS model are presented in Table 6.9. The results obtained by both $\cup$ and $\cap$ operations are further improved by hybridizing the PSO algorithm with the proposed Type-2 FTS model. This new hybridized forecasting model is presented in the next Sect. 6.5.

6.5 Improved Hybridized Forecasting Model

The main downside of FTS forecasting model is that increase in the number of intervals increases accuracy rate of forecasting, and decreases the fuzziness of time series data sets (Singh and Borah 2012). Kuo et al. (2009) show that appropriate selection of intervals also increases the forecasting accuracy of the model. Therefore, in order to get the optimal intervals, they used PSO algorithm in their proposed model (Kuo et al. 2009). Kuo et al. (2009) signify that PSO algorithm is more efficient and powerful than GA (Chen and Chung 2006) in selection of proper intervals. Therefore, to improve the forecasting accuracy of the proposed Type-2 model, we have hybridized the PSO algorithm with Algorithm 5. The main function of the PSO algorithm in Algorithm 5 is to adjust the lengths of intervals and membership values simultaneously, without increasing the number of intervals in the model. We have entitled this model as "FTS-PSO". The detailed description of the FTS-PSO model is presented below.

Let n be the number of intervals, x_0 and x_n be the lower and upper bounds of the universe of discourse U on historical time series data set $D(t)$, respectively. A particle is an array consisting of $n-1$ elements such as $x_1, x_2, \ldots, x_i, \ldots, x_{n-2}$ and x_{n-1}, where $1 \leq i \leq n-1$ and $x_{i-1} < x_i$. Now based on these $n-1$ elements, define the n intervals as $I_1 = (x_0, x_1]$, $I_2 = (x_1, x_2]$, ..., $I_i = (x_{i-1}, x_i]$, ..., $I_{n-1} = (x_{n-2}, x_{n-1}]$ and $I_n = (x_{n-1}, x_n]$, respectively. In case of movement of a particle from one position to another position, the elements of the corresponding new array always require to be adjusted in an ascending order such that $x_1 \leq x_2 \leq \ldots x_{n-1}$. The graphical representation of particle is shown in Fig. 6.1.

In this process, the FTS-PSO model allows the particles to move other positions based on Eqs. 2.4.6 and 2.4.7, and repeats the steps until the stopping criterion is satisfied or the optimal solution is found. If the stopping criterion is satisfied, then employ all the FLRs obtained by the global best position (*gbest*) among all personal

Fig. 6.1 The graphical representation of particle

x_1	x_2	...	x_i	...	x_{n-1}

best positions (*pbest*) of all particles. Here, the AFER is used to evaluate the forecasted accuracy of a particle. The complete steps of the FTS-PSO model is presented in Algorithm 6.

The main difference between the existing models (Kuo et al. 2009, 2010) and the FTS-PSO model is the procedure for handling the intervals based on their importance. The FTS-PSO model also incorporates more information in terms of observations, which are represented in terms of FLRs. These FLRs are later employed for defuzzification based on the technique discussed in Sect. 6.4. In the following, an example is presented to demonstrate the whole process of the FTS-PSO model.

Algorithm 6 FTS-PSO Algorithm

1: Initialize all particles' positions and velocities;
2: **While** (the stopping criterion is not satisfied) **Do**
3: **For** Each particle *id* **Do**
4: Define linguistic terms based on the current position of particle *id*;
5: Fuzzify the time series data set of Type-1 and Type-2 observations according to the linguistic terms defined in the previous step;
6: Establish the FLRs based on Definition 2.2.7;
7: Construct the FLRGs based on Definition 2.2.8;
8: Establish the relationships between FLRGs of both Type-1 and Type-2 observations, and map-out them to their corresponding day;
9: Based on Definitions 6.2.1 and 6.2.2, apply $\cup$ and $\cap$ operators on map-out FLRGs of Type-1 and Type-2 observations, and obtain the fuzzified forecasting data;
10: Defuzzify the forecasting data based on the "FWDT";
11: Calculate the forecasted values individually for $\cup$ and $\cap$ operations;
12: Compute the AFER value for particle *id*;
13: Update the *pbest* and *gbest* for particle *id* according to the AFER;
14: **End For**;
15: **For** Each particle *id* **Do**
16: Move the particle to another position according to Eqs. 2.4.6 and 2.4.7;
17: **End For**;
18: **End**;

[Example] The FTS-PSO model employs the PSO to obtain the optimal FLRs of both Type-1 and Type-2 observations by adjusting the lengths of intervals for the historical data, $D(t)$, where $6/4/2012 \leq t \leq 7/29/2012$ (see Table 6.1). In Sect. 6.4, we have the universe of discourse $U = [1929.50, 2253.10]$, where lower bound $x_0 = 1929.50$ and upper bound $x_{26} = 2253.10$, respectively. On the universe of discourse, total 30 intervals are defined based on Eq. 6.4.1, but the historical data cover only 26 intervals. Therefore, for the representation of particles, we use these 26 intervals. For finding the optimal solution, we define 4 particles. Now, based on $U = [1929.50, 2253.10]$, we define values for the parameters used in Eqs. 2.4.6 and 2.4.7 as: (a) $CP_{id} = [1929.50, 2253.10]$, (b) $Vel_{id,t} = [-3, 3]$, (c) M_1 and $M_2 = 1.5$, and (d) $\alpha = 1.4$ (where α linearly decreases its value to the lower bound, 0.4, through the whole procedure) respectively. The positions and velocities of all particles are initialized randomly and shown in Tables 6.10 and 6.11, respectively.

Table 6.10 Randomly generated initial positions of all particles

Particle	x_1	x_2	x_3	x_4	x_5	x_6	...	x_{25}	AFER (%)
1(∪)	1940.3	1951.1	1961.9	2015.8	2026.6	2037.4	...	2242.3	0.68
1(∩)	1940.3	1951.1	1961.9	2015.8	2026.6	2037.4	...	2242.3	0.66
2(∪)	1938.4	1949.2	1960.1	2014.3	2025.2	2036	...	2242.2	0.66
2(∩)	1938.4	1949.2	1960.1	2014.3	2025.2	2036	...	2242.2	0.67
3(∪)	1932.6	1943.6	1954.7	2009.9	2021.0	2032.0	...	2242.0	0.64
3(∩)	1932.6	1943.6	1954.7	2009.9	2021.0	2032.0	...	2242.0	0.67
4(∪)	1936.4	1947.3	1958.3	2012.8	2023.8	2034.7	...	2242.1	0.67
4(∩)	1936.4	1947.3	1958.3	2012.8	2023.8	2034.7	...	2242.1	0.69

Table 6.11 Randomly generated initial velocities of all particles

Particle	x_1	x_2	x_3	x_4	x_5	x_6	...	x_{25}
1(∪)	2.04	3.70	1.37	3.71	0.01	4.60	...	2.17
1(∩)	2.04	3.70	1.37	3.71	0.01	4.60	...	2.17
2(∪)	4.45	0.15	4.78	0.70	2.63	2.59	...	4.38
2(∩)	4.45	0.15	4.78	0.70	2.63	2.59	...	4.38
3(∪)	2.60	2.45	2.68	2.84	0.43	2.15	...	2.93
3(∩)	2.60	2.45	2.68	2.84	0.43	2.15	...	2.93
4(∪)	3.19	0.07	4.33	1.25	1.69	1.57	...	2.01
4(∩)	3.19	0.07	4.33	1.25	1.69	1.57	...	2.01

In Table 6.10, we have shown the 26 intervals for each particle. For example, the intervals for the initial position of particle 1 are as: $I_1 = (1929.50, 1940.30]$, $I_2 = (1940.30, 1951.10], \ldots, I_{26} = (2242.30, 2253.10]$, respectively. In Table 6.10, we consider those intervals for particle 1 that are used in fuzzification of the time series data set, presented in Sect. 6.4. In this process, we follow the steps of Algorithm 6 (modified version of Algorithm 5), and obtain the optimal FLRs which are employed for obtaining the forecasted results. Then, the AFER value for particle 1 is computed. The AFER values for the remaining 3 particles are obtained in a similar manner. Based on the corresponding AFER value, every particle updates its own *pbest*. For simplicity, the initial *pbests* are considered for the initial positions of all particles (Kuo et al. 2009, 2010). The *pbests* of all particles are shown in Table 6.12. In Table 6.12, the PG_{best} is obtained by particle 3 (for the ∪ operation) and particle 1 (for the ∩ operation).

According to Algorithm 6, all particles move towards the second positions based on Eqs. 2.4.6 and 2.4.7. The second positions for all particles and their corresponding new AFER values are presented in Table 6.13. For example, in Table 6.13, the second

Table 6.12 The initial *pbest* of all particles

Particle	x_1	x_2	x_3	x_4	x_5	x_6	...	x_{25}	AFER (%)
1(∪)	1940.3	1951.1	1961.9	2015.8	2026.6	2037.4	...	2242.3	0.68
1(∩)	1940.3	1951.1	1961.9	2015.8	2026.6	2037.4	...	2242.3	0.66
2(∪)	1938.4	1949.2	1960.1	2014.3	2025.2	2036	...	2242.2	0.66
2(∩)	1938.4	1949.2	1960.1	2014.3	2025.2	2036	...	2242.2	0.67
3(∪)	1932.6	1943.6	1954.7	2009.9	2021.0	2032.0	...	2242.0	0.64
3(∩)	1932.6	1943.6	1954.7	2009.9	2021.0	2032.0	...	2242.0	0.67
4(∪)	1936.4	1947.3	1958.3	2012.8	2023.8	2034.7	...	2242.1	0.67
4(∩)	1936.4	1947.3	1958.3	2012.8	2023.8	2034.7	...	2242.1	0.69

Table 6.13 The second positions of all particles

Particle	x_1	x_2	x_3	x_4	x_5	x_6	...	x_{25}	AFER (%)
1(∪)	1937.3	1948.1	1958.9	2012.8	2023.6	2034.4	...	2239.3	0.65
1(∩)	1937.3	1948.1	1958.9	2012.8	2023.6	2034.4	...	2239.3	0.64
2(∪)	1935.4	1946.2	1957.1	2011.3	2022.2	2033.0	...	2239.2	0.64
2(∩)	1935.4	1946.2	1957.1	2011.3	2022.2	2033.0	...	2239.2	0.66
3(∪)	1929.6	1940.6	1951.7	2006.9	2018.0	2029.0	...	2239.0	0.67
3(∩)	1929.6	1940.6	1951.7	2006.9	2018.0	2029.0	...	2239.0	0.68
4(∪)	1933.4	1944.3	1955.3	2009.8	2020.8	2031.7	...	2239.1	0.63
4(∩)	1933.4	1944.3	1955.3	2009.8	2020.8	2031.7	...	2239.1	0.67

position of particle 1 (for the ∪ operation) is obtained by using Eqs. 6.5.1 and 6.5.2, which are based on Eqs. 2.4.6 and 2.4.7, respectively.

$$\begin{aligned} Vel_{1,1}(\cup) &= 1.4 \times 2.04 + 1.5 \times Rand \times (1940.3 - 1940.3) + 1.5 \times Rand \\ &\quad \times (1932.6 - 1940.3) = -3 \\ Vel_{1,2}(\cup) &= 1.4 \times 3.70 + 1.5 \times Rand \times (1951.1 - 1951.1) + 1.5 \times Rand \\ &\quad \times (1943.6 - 1951.1) = -3 \\ &\vdots \\ Vel_{1,25}(\cup) &= 1.4 \times 2.17 + 1.5 \times Rand \times (2242.3 - 2242.3) + 1.5 \times Rand \\ &\quad \times (2242.0 - 2242.3) = -3 \end{aligned} \tag{6.5.1}$$

Table 6.14 The second *pbest* of all particles

Particle	x_1	x_2	x_3	x_4	x_5	x_6	...	x_{25}	AFER (%)
1(∪)	1937.3	1948.1	1958.9	2012.8	2023.6	2034.4	...	2239.3	0.65
1(∩)	1937.3	1948.1	1958.9	2012.8	2023.6	2034.4	...	2239.3	0.64
2(∪)	1935.4	1946.2	1957.1	2011.3	2022.2	2033.0	...	2239.2	0.64
2(∩)	1935.4	1946.2	1957.1	2011.3	2022.2	2033.0	...	2239.2	0.66
3(∪)	1929.6	1940.6	1951.7	2006.9	2018.0	2029.0	...	2239.0	0.67
3(∩)	1929.6	1940.6	1951.7	2006.9	2018.0	2029.0	...	2239.0	0.68
4(∪)	1933.4	1944.3	1955.3	2009.8	2020.8	2031.7	...	2239.1	0.63
4(∩)	1933.4	1944.3	1955.3	2009.8	2020.8	2031.7	...	2239.1	0.67

$$\begin{aligned} CP_{1,1}(\cup) &= 1940.3 + Vel_{1,1}(\cup) = 1940.3 + (-3) = 1937.3 \\ CP_{1,2}(\cup) &= 1951.1 + Vel_{1,2}(\cup) = 1951.1 + (-3) = 1948.1 \\ &\vdots \\ CP_{1,25}(\cup) &= 2242.3 + Vel_{1,25}(\cup) = 2242.3 + (-3) = 2239.3 \end{aligned} \quad (6.5.2)$$

On comparison of the AFER values between Tables 6.12 and 6.13, it is obvious that particle 1, particle 2 and particle 4 for the ∪ and ∩ operations attained their own *pbest* values so far in Table 6.13. Therefore, these three particles update their *pbest* values, which are shown in Table 6.14. In Table 6.14, the new PG_{best} value is obtained by particle 4 for the ∪ operation and particle 1 for the ∩ operation, because their AFER values are least among all the particles so far.

The above steps are repeated by the FTS-PSO model until the optimal solution is found or the maximal moving steps are reached. After execution, a new set of optimal FLRs are obtained by the PG_{best} that the particle 4 (∪ operation) and particle 1 (∩ operation) attain so far, and are further employed for obtaining the final forecasting results.

6.6 Empirical Analysis

To illustrate the forecasting performance of the proposed method, the daily stock index price of SBI and Google are used as data sets in verification and validation phases, respectively. The experimental results of the proposed model are compared with different existing models for various orders of FLRs and different intervals.

Table 6.15 A comparison of the forecasted accuracy for the FTS-PSO model and the existing models based on the first-order FLRs

Model	AFER (%)
Chen (1996)	1.34
Yu (2005)	1.29
FTS-PSO (∪ operation)	0.63
FTS-PSO (∩ operation)	0.64

6.6.1 *Stock Index Price Forecasting of SBI*

In this subsection, we present the forecasting results of the FTS-PSO model. The FTS-PSO model is validated using the stock index data set of SBI, as mentioned in Sect. 6.4. For forecasting the stock index price, "Open", "High" and "Low" variables are considered as the Type-2 observations, whereas "Close" variable is considered as the Type-1 observation. The "Actual Price" is chosen as the main forecasting objective. Further comparisons on the FTS-PSO model and the other existing models are discussed below.

The FTS-PSO model is trained simultaneously for the ∪ and ∩ operations, and the best results obtained by the particles are considered to forecast the stock index data. The necessary setting of all the parameters for the FTS-PSO model is discussed in Sect. 6.5.

The best forecasted accuracies (i.e., the least AFER) are made by particle 4 (for the ∪ operation) and particle 1 (for the ∩ operation). Therefore, results obtained by these particles are used for the empirical analysis. The forecasted results for the ∪ and ∩ operations are presented in Table 6.15. In Table 6.15, the forecasted results for the existing FTS models (Chen 1996; Yu 2005) are also presented. The considered FTS models including the FTS-PSO model use the first-order FLRs to forecast the stock index data. From Table 6.15, it is obvious that the FTS-PSO model is more advantageous than the conventional FTS models (Chen 1996; Yu 2005).

To verify the superiority of the proposed model under various high-order conditions, existing forecasting model, *viz.*, Chen model (Chen 2002), is selected for comparison. A comparison of the forecasted results is shown in Table 6.16. During simulation, the number of intervals is kept fix (i.e., 9) for the existing model and the FTS-PSO model. At the same intervals, the AFER values obtained for the existing model are 1.39, 1.41, 1.40, 1.41, 1.40 and 1.40 % for third-order, fourth-order, fifth-order, sixth-order, seventh-order and eight-order FLRs, respectively. On the other

Table 6.16 A comparison of the forecasted accuracy (in terms of AFER) between the FTS-PSO model and the high-order model with different orders (the number of intervals = 9)

Model	Order						FTS-PSO	FTS-PSO
	3	4	5	6	7	8	(∪ operation)	(∩ operation)
Chen (2002)	1.39 %	1.41 %	1.40 %	1.41 %	1.40 %	1.40 %	1.07 %	1.21 %

Table 6.17 A comparison of the forecasted accuracy (in terms of the AFER) for the FTS-PSO model based on the different number of intervals

Model	Number of intervals						
	9	10	11	12	13	14	15
FTS-PSO (∪ operation)	1.07 %	1.04 %	1.20 %	0.88 %	0.89 %	0.91 %	0.82 %
FTS-PSO (∩ operation)	1.21 %	0.99 %	1.22 %	1.09 %	1.06 %	1.04 %	1.00 %

hand, at the same intervals, the FTS-PSO model gets the least AFER values which are 1.07 % (for ∪ operation) and 1.21 % (for ∩ operation). However, the smallest AFER value, which is 1.07 %, is obtained from the proposed model for the ∪ operation. We can see that the FTS-PSO model outperforms the existing models under various high-order FLRs at all.

To verify the performance of the proposed model under different number of intervals, the forecasted results are obtained with different intervals ranging from 9 to 15. The forecasted results are listed in Table 6.17, where the proposed model uses first-order FLRs under different number of intervals. The least AFER values are 0.82 % (for ∪ operation) and 0.99 % (for ∩ operation) for the intervals 15 and 10, respectively. However, between intervals 9 and 15, the best forecasted result is obtained from the ∪ operation at interval 15 (AFER = 0.82 %).

To evaluate the performance of the proposed model, it is compared with the existing Type-2 FTS model (Huarng and Yu 2005), under different number of intervals. A comparison of the forecasted results between the proposed model and the existing Type-2 model is shown in Table 6.18, where both these models use different intervals ranging from 16 to 20. For the existing model, the lowest forecasting error is 1.13 %, which is obtained at intervals 19 and 20. For the proposed model, the least AFER values are 0.72 % (for ∪ operation) and 0.78 % (for ∩ operation), which are obtained at intervals 20 and 19, respectively. From comparison, it is obvious that the proposed model produces more precise results than existing Type-2 model under different number of intervals.

To verify the superiority of the proposed model in terms of forecasted accuracy, three existing models, *viz.*, Grey model (Samvedi and Jain 2013), BPNN (Singh and Borah 2013b) (with one hidden layer and one output layer) and Hybridized model

Table 6.18 A comparison of the forecasted accuracy (in terms of the AFER) between the FTS-PSO model and the existing Type-2 model based on the different number of intervals

Model	Number of intervals				
	16	17	18	19	20
Type-2 model (Huarng and Yu 2005)	1.15 %	1.15 %	1.18 %	1.13 %	1.13 %
FTS-PSO (∪ operation)	0.88 %	0.85 %	0.79 %	0.75 %	0.72 %
FTS-PSO (∩ operation)	0.91 %	0.93 %	0.80 %	0.78 %	0.81 %

based on FTS and ANN (Singh and Borah 2013a), are selected for comparison. These three competing models are simulated using MATLAB (version 7.2.0.232 (R2006a)). During the learning process of the BPNN, a number of experiments were carried out to set additional parameters, *viz.*, *initial weight*, *learning rate*, *momentum* and *minimum weight delta* to obtain the optimal results, and we have chosen the ones that exhibit the best behavior in terms of accuracy. In this work, the *initial weight*, *learning rate*, *momentum* and *minimum weight delta* are taken as: 0.3, 0.5, 0.6 and 0.0001, respectively. The forecasted results for these three models are obtained with different number of input values ranging from 5 to 10. A comparison of the forecasted results is presented in Table 6.19. The least AFER values for the Grey model, BPNN model and Hybridized model are 1.05 % (for input 5), 1.25 % (for input 10) and 1.17 % (for input 10), respectively. In our proposed model, the selection of input depends on the establishment of FLRs. For the proposed model, the least AFER values are 0.63 % (for $\cup$ operation) and 0.64 % (for $\cap$ operation), which are obtained using the first-order FLRs. From comparison, we can see that the proposed model gets a higher forecasting accuracy than the existing three models, *viz.*, Grey model, BPNN model and Hybridized model.

The empirical analysis shows that the proposed model is far better than the existing forecasting models for stock index data set of SBI.

6.6.2 Stock Index Price Forecasting of Google

In this subsection, the performance of the proposed model is evaluated with the stock index data set of Google. The data set of stock index price is covered from the period 6/1/2012 − 7/27/2012, which is shown in Table 6.20. Here, "Open", "High" and "Low" variables are selected as the Type-2 observations, whereas "Close" variable is selected as the Type-1 observation. The "Actual Price" is chosen as the main forecasting objective. The historical stock index data set of Google is collected from the website.[4]

The FTS-PSO model optimizes the forecasting results and obtains the best results (i.e., the least AFER). The forecasting results are shown in Table 6.21. The universe of discourse U is considered as $U = [555, 637]$, and is partitioned into 30 intervals. But, the historical data cover only 24 intervals, and the proposed model obtains the forecasted results using these 24 intervals. More detail results on partitions of the universe of discourse and positions of the particle (i.e., best particle) among these 24 intervals are shown in Table 6.22.

For comparison studies, various statistical models listed in article (Aladag et al. 2010) are simulated using PASW Statistics 18.[5] A comparison of the forecasted accuracy among the conventional statistical models and the proposed model is listed

[4]http://in.finance.yahoo.com.

[5]http://www.spss.com.hk/statistics/.

Table 6.19 A comparison of the forecasted accuracy (in terms of AFER) between the FTS-PSO model and the existing models (with different number of input values)

Model	Input						FTS-PSO	FTS-PSO
	5	6	7	8	9	10	(∪ operation)	(∩ operation)
Grey model (Samvedi and Jain 2013)	1.05 %	1.66 %	1.62 %	1.95 %	2.34 %	2.69 %	0.63 %	0.64 %
BPNN model (Singh and Borah 2013b)	1.33 %	1.47 %	1.53 %	1.51 %	1.58 %	1.25 %	–	–
Hybridized model (Singh and Borah 2013a)	1.23 %	1.21 %	1.21 %	1.21 %	1.25 %	1.17 %	–	–

Table 6.20 Daily stock index price list of Google (in USD)

Date (mm/dd/yy)	Open	High	Low	Close	Actual price
6/1/2012	571.79	572.65	568.35	570.98	570.94
6/4/2012	570.22	580.49	570.01	578.59	574.83
6/5/2012	575.45	578.13	566.47	570.41	572.62
6/6/2012	576.48	581.97	573.61	580.57	578.16
6/7/2012	587.60	587.89	577.25	578.23	582.74
6/8/2012	575.85	581.00	574.58	580.45	577.97
6/11/2012	584.21	585.32	566.69	568.50	576.18
...	...	...	...	...	...
7/27/2012	618.89	635.00	617.50	634.96	626.59

Table 6.21 Forecasted results of the stock index price of Google (in USD)

Date (mm/dd/yy)	Actual price	FTS-PSO ($\cup$ operation)	FTS-PSO ($\cap$ operation)
6/1/2012	570.94	574.32	568.01
6/4/2012	574.83	577.21	572.43
6/5/2012	572.62	576.28	569.92
6/6/2012	578.16	580.6	572.68
6/7/2012	582.74	585.32	579.19
6/8/2012	577.97	583.7	574.29
6/11/2012	576.18	574.38	570.32
...	...	...	...
7/27/2012	626.59	622.31	622.31

Table 6.22 Partitions of the universe of discourse and positions of the particle (i.e., best particle) among these intervals

Model	x_1	x_2	x_3	x_4	x_5	...	x_{23}	AFER (%)
FTS-PSO ($\cup$ operation)	557.73	560.47	563.20	565.93	568.67	...	620.60	0.6733
FTS-PSO ($\cap$ operation)	557.73	560.47	563.20	565.93	568.67	...	620.60	0.4549

in Table 6.23. The comparative analysis clearly shows that the proposed model outperforms the considered models.

[Robustness] To check the robustness of the proposed model for forecasting the stock index price of Google, various statistical parameters values as mentioned in Sect. 2.6 (see Chap. 2), are obtained. The experimental results are shown in Table 6.24. The

Table 6.23 A comparison of the forecasted accuracy (in terms of AFER) between the FTS-PSO model and the existing statistical models

Model	AFER (%)
Logarithmic regression	1.9970
Inverse regression	2.0472
Quadratic regression	1.2683
Cubic regression	1.2212
Compound regression	1.7123
Power regression	1.9891
S-curve regression	2.0370
Growth regression	1.7123
Exponential regression	0.8590
FTS-PSO (∪ operation)	0.6733
FTS-PSO (∩ operation)	0.4549

Table 6.24 Statistics of the FTS-PSO model for the stock index price forecasting of Google

Statistics	Actual price	Forecasted price (∪ operation)	Forecasted price (∩ operation)
Mean (USD)	580.34	583.43	578.11
SD (USD)	16.12	14.99	16.08
R (USD)	–	0.98	0.99
U (USD)	–	0.0038	0.0028

values of parameters listed in Table 6.24 are based on the forecasted results presented in Table 6.21.

From Table 6.24, it is clear that the mean of actual price is very close to the mean of forecasted price. The comparison of the SD values between actual price and forecasted price show that predictive skill of our proposed model is good for both the ∪ and ∩ operations. The R values between actual and forecasted values also indicate the efficiency of the proposed model. The U values for both ∪ and ∩ operations are closer to 0, which indicate the effectiveness of the proposed model. Hence, the robustness of the proposed model is strongly convinced with the outstanding performance in case of daily stock index price data set of Google.

6.7 Discussion

This chapter presents an approach combining Type-2 FTS with the PSO for building a time series forecasting expert system. The main contributions of this chapter are presented as follows:

- **First**, the author shows that the problem of stock index price forecasting can be solved using Type-2 FTS concept. In this work, the author demonstrates the application of the Type-2 model on 4-factors (i.e., "Low", "Medium", "High" and "Close") time series data set of SBI.
- **Second**, the author shows how the assignment of weights on intervals based on their frequencies and later utilization of these frequencies in defuzzification process, can improve the forecasting accuracy of the proposed model.
- **Third**, the author shows that the accuracy rate of the stock index price forecasting can be improved effectively by hybridizing the PSO algorithm with the Type-2 model.
- **Fourth**, the author shows that the forecasting accuracy of the FTS-PSO model is more precise than the existing FTS models.
- **Fifth**, the author shows the robustness of the FTS-PSO model by comparing its forecasting accuracy with various statistical models.

Still, there are scopes to apply the model in some other domains in a flexible way as follows:

1. To check the accuracy and performance of the model by forecasting the weather for different regions, and
2. To test the performance of the model for different types of financial, stocks and marketing data set.

References

Aladag CH (2013) Using multiplicative neuron model to establish fuzzy logic relationships. Expert Syst Appl 40(3):850–853

Aladag CH, Yolcu U, Egrioglu E (2010) A high order fuzzy time series forecasting model based on adaptive expectation and artificial neural networks. Math Comput Simul 81(4):875–882

Almaraashi M, John R (2010) Tuning fuzzy systems by simulated annealing to predict time series with added noise. Proceedings of UKCI. Essex, UK, pp 1–5

Almaraashi M, John R (2011) Tuning of type-2 fuzzy systems by simulated annealing to predict time series. In: Proceedings of the international conference of computational intelligence and intelligent systems, London

Chang JR, Wei LY, Cheng CH (2011) A hybrid ANFIS model based on AR and volatility for TAIEX forecasting. Appl Soft Comput 11(1):1388–1395

Chen SM (1996) Forecasting enrollments based on fuzzy time series. Fuzzy Sets Syst 81:311–319

Chen SM (2002) Forecasting enrollments based on high-order fuzzy time series. Cybern Syst Int J 33(1):1–16

Chen SM, Chung NY (2006) Forecasting enrollments of students by using fuzzy time series and genetic algorithms. Int J Inf ManageSci 17(3):1–17

Cheng CH, Wei LY, Liu JW, Chen TL (2013) OWA-based ANFIS model for TAIEX forecasting. Econ Model 30:442–448

Hsu LY, Horng SJ, Kao TW, Chen YH, Run RS, Chen RJ, Lai JL, Kuo IH (2010) Temperature prediction and TAIFEX forecasting based on fuzzy relationships and MTPSO techniques. Expert Syst Appl 37(4):2756–2770

Huarng K, Yu HK (2005) A Type 2 fuzzy time series model for stock index forecasting. Phys A Stat Mech Appl 353:445–462

Karnik NN, Mendel JM, Liang Q (1999) Type-2 fuzzy logic systems. IEEE Trans Fuzzy Syst 7(6):643–658

Kumbasar T, Eksin I, Guzelkaya M, Yesil E (2013) Exact inversion of decomposable interval type-2 fuzzy logic systems. Int J Approximate Reasoning 54(2):253–272

Kuo IH, Horng SJ, Kao TW, Lin TL, Lee CL, Pan Y (2009) An improved method for forecasting enrollments based on fuzzy time series and particle swarm optimization. Expert Syst Appl 36(3, Part 2):6108–6117

Kuo IH, Horng SJ, Chen YH, Run RS, Kao TW, Chen RJ, Lai JL, Lin TL (2010) Forecasting TAIFEX based on fuzzy time series and particle swarm optimization. Expert Syst Appl 37(2):1494–1502

Lee HS, Chou MT (2004) Fuzzy forecasting based on fuzzy time series. Int J Comput Math 81(7):781–789

Liu HT, Wei ML (2010) An improved fuzzy forecasting method for seasonal time series. Expert Syst Appl 37(9):6310–6318

Mencattini A, Salmeri M, Bertazzoni S, Lojacono R, Pasero E, Moniaci W (2006) Short term local meteorological forecasting using type-2 fuzzy systems. Neural Nets Lect Notes Comput Sci 3931:95–104

Molaeezadeh SF, Moradi MH (2013) A 2ufunction representation for non-uniform type-2 fuzzy sets: theory and design. Int J Approximate Reasoning 54(2):273–289

Park J, Lee DJ, Song CK, Chun MG (2010) TAIFEX and KOSPI 200 forecasting based on two-factors high-order fuzzy time series and particle swarm optimization. Expert Syst Appl 37(2):959–967

Pedrycz W, Song M (2012) Granular fuzzy models: a study in knowledge management in fuzzy modeling. Int J Approximate Reasoning 53(7):1061–1079

Ross TJ (2007) Fuzzy Logic with engineering applications. John Wiley and Sons, Singapore

Samvedi A, Jain V (2013) A grey approach for forecasting in a supply chain during intermittent disruptions. Engineering Applications of Artificial Intelligence 26(3):1044–1051

Singh P, Borah B (2012) An effective neural network and fuzzy time series-based hybridized model to handle forecasting problems of two factors. Knowl Inf Syst 38(3):669–690

Singh P, Borah B (2013a) High-order fuzzy-neuro expert system for daily temperature forecasting. Knowl Based Syst 46:12–21

Singh P, Borah B (2013b) Indian summer monsoon rainfall prediction using artificial neural network. Stoch Environ Res Risk Assess 27(7):1585–1599

Wei LY (2013) A GA-weighted ANFIS model based on multiple stock market volatility causality for TAIEX forecasting. Appl Soft Comput 13(2):911–920

Yu HK (2005) Weighted fuzzy time series models for TAIEX forecasting. Phys A Stat Mech Appl 349(3–4):609–624

Chapter 7
Indian Summer Monsoon Rainfall Prediction

I still believe in the possibility of a model of reality, that is to say, of a theory, which represents things themselves and not merely the probability of their occurrence.

By Einstein (1879–1955)

Abstract Forecasting the monsoon temporally is a major scientific issue in the field of monsoon meteorology. The ensemble of statistics and mathematics has increased the accuracy of forecasting of ISMR up to some extent. But due to the nonlinear nature of ISMR, its forecasting accuracy is still below the satisfactory level. Mathematical and statistical models require complex computing power. Therefore, many researchers have paid attention to apply ANN in ISMR forecasting. In this study, we have used Feed-Forward Back-Propagation neural network algorithm for ISMR forecasting. Based on this algorithm, we have proposed the five neural network architectures designated as BP1, BP2, ..., BP5 using three layers of neurons (one input layer, one hidden layer and one output layer). Detail architecture of the neural networks are are provided in this chapter. Time series data set of ISMR is obtained from Pathasarathy (1994) (1871–1994) and IITM (2012) (1995–2010) for the period 1871–2010, for the months of June, July, August and September individually, and for the monsoon season (sum of June, July, August and September). The data set is trained and tested separately for each of the neural network architecture, *viz.*, BP1–BP5. The forecasted results obtained for the training and testing data are then compared with existing model. Results clearly exhibit superiority of our model over the considered existing model. The seasonal rainfall values over India for next 5 years have also been predicted.

Keywords ISMR · FFNN · BBNN · Time series · Drought

P. Singh, *Applications of Soft Computing in Time Series Forecasting*,
Studies in Fuzziness and Soft Computing 330, DOI 10.1007/978-3-319-26293-2_7

7.1 Background and Related Literature

The Indian economy is based on agriculture and its products, and crop yield is heavily dependent on the summer monsoon (June–September) rainfall. Therefore, any decrease or increase in annual rainfall will always have a severe impact on the agricultural sector in India. About 65 % of the total cultivated land in India is under the influence of rain-fed agriculture system (Swaminathan 1998). Therefore, prior knowledge of the monsoon behavior (during which the maximum rainfall occurs in a concentrated period) will help the Indian farmers and the Government to take advantage of the monsoon season. This knowledge can be very useful in reducing the damage of crops during less rainfall periods in monsoon season. Therefore, forecasting the monsoon temporally is a major scientific issue in the field of monsoon meteorology.

The ensemble of statistics and mathematics has increased the accuracy of forecasting of ISMR up to some extent. But due to the non-linear nature of ISMR, its forecasting accuracy is still below the satisfactory level. In 2002, IMD failed to predict the deficit of rainfall during ISMR, which led to considerable concern in the meteorological community (Gadgil et al. 2002). In 2004, drought was again observed in the country with a deficit of more than 13 % rainfall (Gadgil et al. 2005), which could not be predicted by any statistical or dynamic model. Preethi et al. (2011) reported that India as a whole received 77 % of rainfall during ISMR in 2009, which was the third highest deficient of all ISMR years during the period 1901–2009.

Various experiments were done by researchers to recognize the suitable prediction parameters for forecasting ISMR. Forecasting of ISMR started more than 100 years ago (Blanford 1884; Walker 1933). Mathematical and statistical models require complex computing power (Krishna et al. 1995; Mooley and Parthasarathy 1984; Satyan 1988; Basu and Andharia 1992). Therefore, many researchers tried to apply ANN for ISMR forecasting. In literature, several types of neural networks can be found. But usually, only FFNN and BPNN are used in ISMR forecasting.

Goswami and Srividya (1996) forecasted the annual mean rainfall fifteen years in advance with an average error rate of less than 10 %. They proposed a new neural network model called CN, which is trained by the BPNN algorithm. The experimental results show that the CN is better than the conventional ANN model. Navone and Ceccatto (1994) forecasted the ISMR using the FFNN. In this approach, ISMR predictors as used in these articles (Shukla and Mooley 1987; Hastenrath 1988), are correlated with the rainfall anomaly using neural network rather than the multiple linear regression equation. The experimental results show that the performance of the ANN approach is better than conventional approaches. Guhathakurta et al. (1999) developed three different types of models for long-range prediction of ISMR as a discrete input layer model (first model), principal Component model (second model), and hybridization of first and second model (third model) with the help of two layer hybrid neural network. Sahai et al. (2000) predicted the seasonal and monthly mean rainfall over India using FFNN with EBP, and reported that ISMR has scale variability in predictability and is independent of any teleconnection.

Guhathakurta (2006) suggested that ANN models are better than the statistical models which are mostly used by IMD. He forecasted the rainfall during monsoon season for districts of Kerala (India) using the FFNN with back-propagation learning algorithm. Chakraverty and Gupta (2007) predicted the southwest monsoon rainfall in India for 6 years in advance. They used the EBP ANN algorithm along with the supervised learning method, and the experimental results show that their proposed model is better than many existing models.[1] Aksoy and Dahamsheh (2008) forecasted the precipitation for 1-month advance using the BBNN, RBF and GR neural networks. Later, all these three types of ANN models are compared with MLR, and BBNN is reported to be better than RBF and GR neural network including MLR. But in case of low precipitation region, RBF is better than BBNN.

7.2 Description of Data Sets

In India, ISMR starts in the month of June and ends in the month of September. July and August fall in the mid of the monsoon season. In this work, we select monthly (June, July, August and September) and seasonal (sum of June, July, August and September) rainfall for all India as the main forecasting objective. India (as a whole) is a large country and due to the high spatial variability of monsoon rainfall over India, it is unusual to find some areas with deficient rain even with the best performance of the summer monsoon, and some areas of floods even with the worst performance of the summer monsoon season. It is very complex to incorporate all these seasonal variabilities in a single series. Therefore, for various climatological studies, Mooley and Parthasarathy (1984) suggested to take the arithmetic average of the rainfall values of the stations over the region, which also help in reducing the climatic noises or missing values present in data (especially daily data, which contain lots of noises and missing values). Hence, Parthasarathy et al. (1992, 1994) prepared all-India average monthly rainfall values by weighing each of the subdivisional rainfall area (306 well-distributed rain-guages). These monthly data can be obtained from Parthasarathy et al. (1994) (1871–1994) and IITM[2] (1995–2010) for the period 1871–2010, for each of the individual month and seasonal.

In this work, the time series data for 140 years (1871–2010) are divided into two parts as: (a) Training set from the period 1871–1960, and (b) Testing set from the period 1961–2010. Thus, there are ($140 \times 5 =$) 700 entries of rainfall values in our model. As the previous year's rainfall values are used for forecasting the next year, therefore predictions are available in the training set from the period 1876–1960 and in the testing set from the period 1961–2010.

---

[1]References are: (Rajeevan et al. 2004; Goswami and Srividya 1996; Goswami and Kumar 1997).

[2]http://www.tropmet.res.in/,2012.

7.3 Descriptive Statistics

The Pearson's correlation values between the four months (June to September) are depicted in Table 7.1, which reflect that rainfall is not pair-wise correlated. The correlation values for pair June–July (−0.0342), June–August (−0.0311), June–September (−0.0695), July–August (0.1005), July–September (0.2799) and August–September (0.2444) are very small, which also suggested that the relationships are not linear. To analyze the relationship of ISMR with its past and future values, ACFs (Chattopadhyay and Chattopadhyay 2001) are obtained for each month and seasonal rainfall. The curves of ACFs for the four months and seasonal time series are depicted in Figs. 7.1, 7.2, 7.3, 7.4 and 7.5. In these figures, autocorrelation coefficients of rainfall for each of the time series vary from +0.5 to −0.5 (corresponding to lags $1, 2, \ldots, 100$). This indicates that ISMR time series for each of the four months and seasonal exhibit *no persistence*.

Table 7.1 Correlation analyzes of ISMR data for the period 1871–2010

Correlation	June	July	August	September
June	1	−0.0342	−0.0311	−0.0645
July	−0.0342	1	0.1005	0.2799
August	−0.0311	0.1005	1	0.2444
September	−0.0645	0.2799	0.2444	1

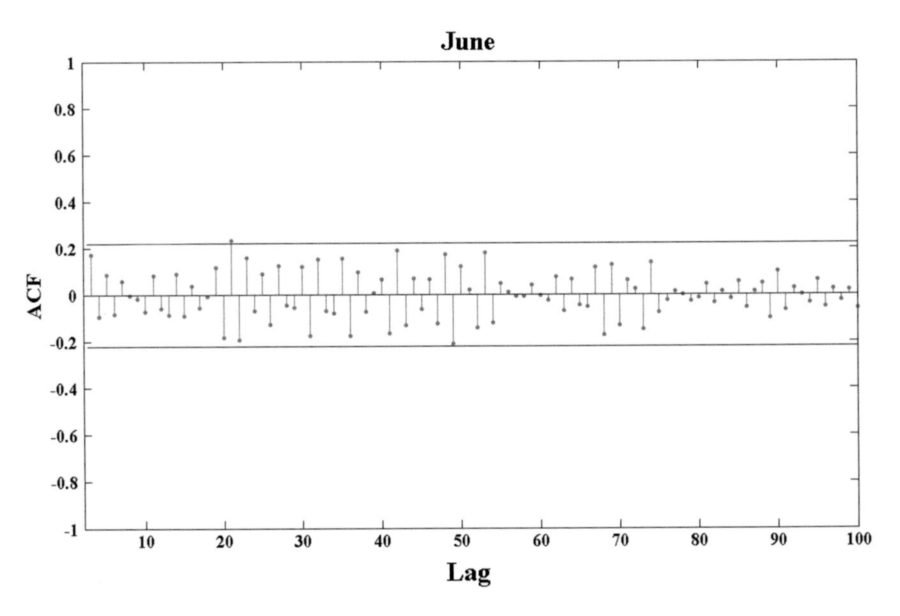

Fig. 7.1 Curve showing the ACF for June rainfall values (1871–2010)

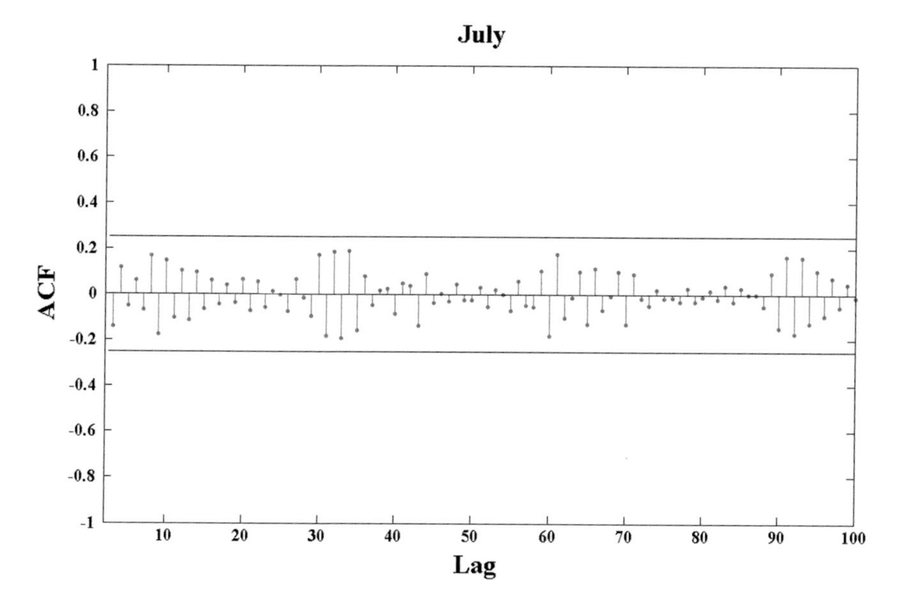

Fig. 7.2 Curve showing the ACF for July rainfall values (1871–2010)

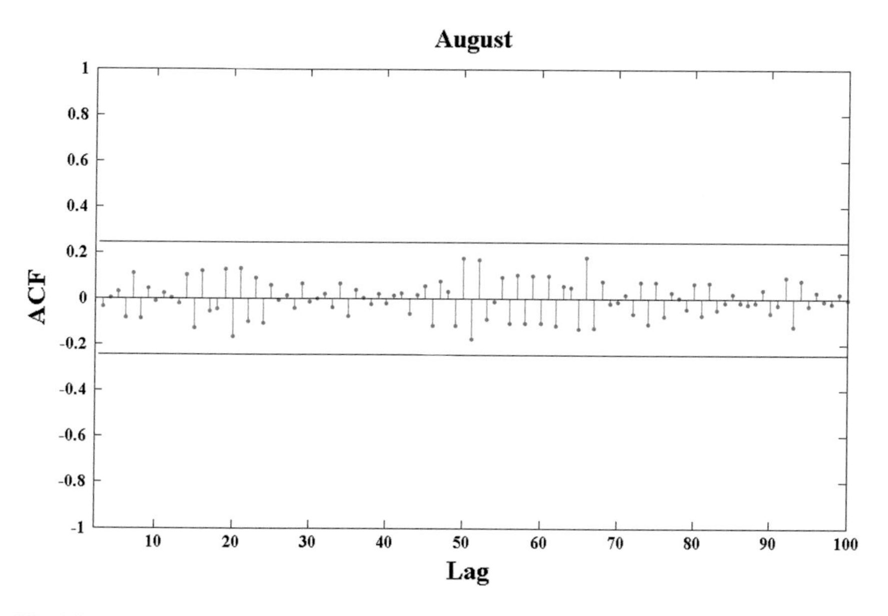

Fig. 7.3 Curve showing the ACF for August rainfall values (1871–2010)

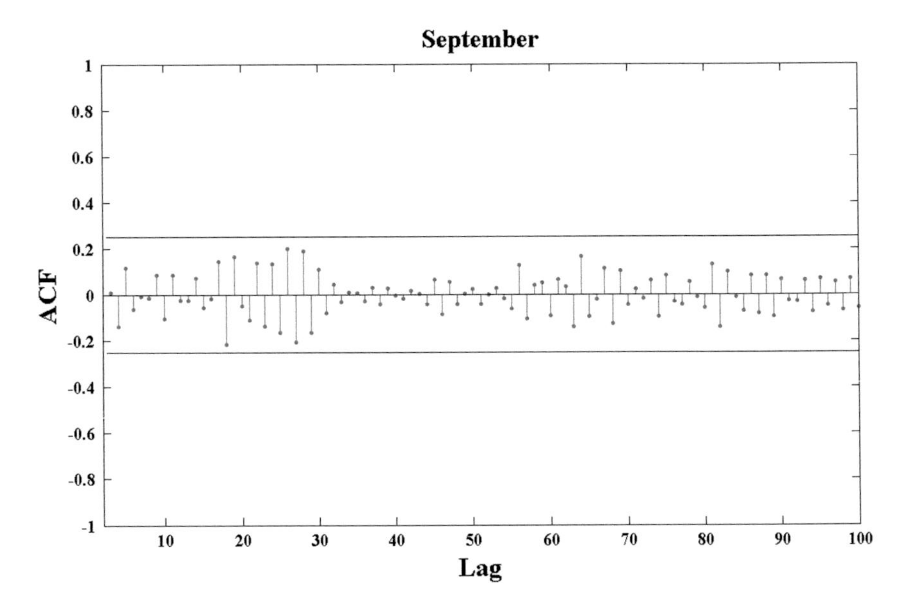

Fig. 7.4 Curve showing the ACF for September rainfall values (1871–2010)

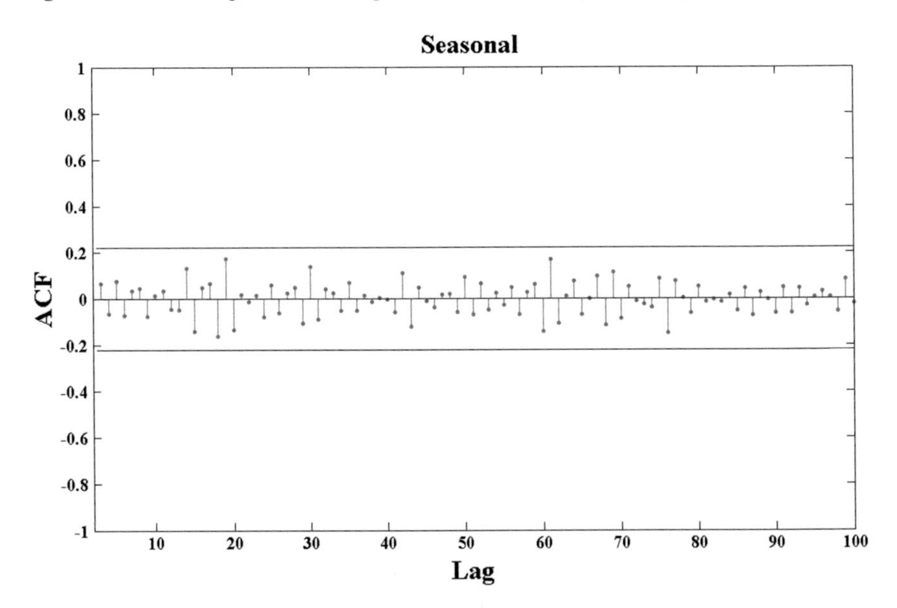

Fig. 7.5 Curve showing the ACF for Seasonal rainfall values (1871–2010)

Table 7.2 Statistical summary of ISMR data for the period 1871–2010

Statistics	June	July	August	September	Seasonal
Mean (mm)	163.88	272.22	241.85	170.22	848.17
Min (mm)	78.2	117.6	144.1	77.2	603.9
Max (mm)	241.6	346	339.3	267.8	1020.1
St. dev. (mm)	36.59	37.98	37.91	37.12	83.70
Skewness	−0.08	−1.19	0.04	0.12	−0.53
Kurtosis	2.39	5.56	2.56	2.42	2.96

Normality of the distribution of each time series is checked by analyzing skewness and kurtosis test (Johnson and Wichern 1992; MATLAB 2006). A complete statistical summary of the distribution of ISMR data for the period 1871–2010 is presented in Table 7.2. The time series data for the months June–September demonstrated the skewness values ranging from −0.08 to 0.12, and kurtosis values ranging from 2.56 to 5.56. This indicates that distributions of rainfall throughout the monsoon season are far from normal with 95 % confidence.

All these statistical results imply the significance of designing an ANN based model for advance prediction of rainfall. Therefore, in this work, an ANN based model is presented to predict ISMR of a given year using the observed time series data of the four months and seasonal. The model is developed based on supervised BPNN algorithm where the learning process aims to minimize the error rate between predicted output and the actual observation. The performance of the model is assessed using various statistical parameters. Large number of input patterns may lead to overfitting of a model, and the determination of suitable predictor parameters for ISMR is not yet possible as far as existing literatures are concerned.[3] So, we try to predict ISMR based on the monthly time series rainfall values. In this work, an attempt has also been made to predict the seasonal rainfall amounts for 5 years in advance.

7.4 Description of the Neural Network Based Method

7.4.1 Architecture of the Proposed Model

The architecture of multi-layer neural network is more complex than single-layer neural network (as discussed in Chap. 2). In the proposed neural network architecture, some additional complexities are removed by considering the following paradigms:

[3]References are:(Krishnamurti et al. 1990; Hastenrath and Greischar 1993; Annamalai 1995; Sahai et al. 2000).

1. A MLFF neural network with a nonlinear activation function can classify the data very efficiently (Pandya and Macy 1996). Therefore, this type of neural network is considered in the development of our architecture.
2. A MLFF neural network with more than three layers can generate arbitrarily complex decision regions (Lippmann 1989). Therefore, a single hidden layer with one input layer and one output layer is considered in designing the architecture.
3. A large number of neurons in hidden layer can make the training process of MLFF neural network more complex, because the weight of each interconnection link needs to be adjusted in each iteration of the training process. Therefore, the proposed neural network is designed with minimum number of neurons in hidden layer. During the training process, each time an error is calculated while adjusting the weights. To minimize this error, BPNN is used, where the error is propagated back to the hidden layers (Masters 1993; Bishop 1994). Therefore, BPNN is integrated with MLFF neural network in our architecture.

Based on the mentioned paradigms, we have proposed the five neural network architectures designated as BP1, BP2, ..., BP5 using three-layers of neurons (one input layer, one hidden layer and one output layer). In this work, the minimum number of neurons in the input and hidden layers are determined by the following equation (Carpenter and Barthelemy 1994):

$$HL_{nodes} = IL_{nodes} + 1, \tag{7.4.1}$$

where HL_{nodes} and IL_{nodes} represent the number of neurons in the hidden and input layers, respectively. The description of the neural network architectures with different HL_{nodes} and IL_{nodes} are presented in Table 7.3. By subsequently increasing the number of nodes in the input and hidden layers in the architecture of BP1 neural network, the rest of the neural network architectures as listed in Table 7.3, can be obtained.

To control the training process of the neural network, initial weights, learning rate, momentum, epoch, minimum weight delta, activation function, etc. parameters are used. The detailed description of all these parameters can be found in Sivanandam and Deepa (2007).

Table 7.3 Description of the five neural networks

Designation	IL_{nodes}	HL_{nodes}	Output layer node
BP1	6	7	1
BP2	7	8	1
BP3	8	9	1
BP4	9	10	1
BP5	10	11	1

7.4.2 Learning Process of Neural Networks

To explain the learning process of neural networks, only BP1 neural network is considered here as an example. In this neural network, there are 6 nodes in input layer. The arrangement of patterns in the input layer is done in the following sequence:

$$I_{jun} = [(X_i^{t-4}, X_i^{t-3}, X_i^{t-2}, X_i^{t-1}, X_i^t)], \tag{7.4.2}$$

$$I_{jul} = [(X_j^{t-4}, X_j^{t-3}, X_j^{t-2}, X_j^{t-1}, X_j^t)], \tag{7.4.3}$$

$$I_{aug} = [(X_k^{t-4}, X_k^{t-3}, X_k^{t-2}, X_k^{t-1}, X_k^t)], \tag{7.4.4}$$

$$I_{sep} = [(X_l^{t-4}, X_l^{t-3}, X_l^{t-2}, X_l^{t-1}, X_l^t)], \tag{7.4.5}$$

$$I_{seas} = [(X_m^{t-4}, X_m^{t-3}, X_m^{t-2}, X_m^{t-1}, X_m^t)], \tag{7.4.6}$$

where X_i, X_j, X_k, X_l and X_m denote the recorded rainfall values of June (I_{jun}), July (I_{jul}), August (I_{aug}), September (I_{sep}) and seasonal (I_{seas}), respectively. Here, each "t" represents the year, i.e., "$t-4$" denotes 1871, "$t-3$" denotes 1872 and so on.

BP1 neural network is trained individually for the five time series data of June, July, August, September and seasonal. The prediction of one year rainfall (i.e., rainfall values of $(t-4)$, $(t-3)$, $(t-2)$, $\ldots$, t years are used to predict $(t+1)$ year rainfall value) is obtained from the previous five years rainfall values of the training data (1871–1960). Therefore, there is a five year overlap in the prediction of rainfall values for BP1 neural network, and thus predictions are available in the training set from the period 1876–1960. Similarly, for BP2–BP5 neural networks, predictions are available in the training set from the periods 1877–1960, 1878–1960, 1879–1960 and 1880–1960, respectively. The phases employed for the training process of BP1 neural network are presented in **Appendix A** of the article (Singh and Borah 2013), written by Singh and Borah. Remaining neural networks are trained in a similar manner.

In the present approach, the weights are required to be adjusted in each iteration of the training phases, which leads to an increase in training time of the architecture. So, to over come this problem, the weights are updated after all the training information are presented. Further, this neural network is carried out for testing process. The phases employed for the testing process of the neural network is presented in **Appendix B** of the article (Singh and Borah 2013), written by Singh and Borah.

A convergence problem occurs if the original data is used as the input to the neural network (Khan and Ondrusek 2000; Maqsood et al. 2004). To avoid this problem, the scaling of data is done using z-score normalization (Han and Kamber 2001; Wu et al. 2008). For example, the elements of time series data "I_i" are normalized based on the mean and standard deviation of "I_i". An element "X" of "I_i" is normalized to "$\grave{X}$" by computing:

$$\grave{X} = \frac{X - \bar{I}_i}{\sigma_{I_i}}, \tag{7.4.7}$$

where $\bar{I}_i$ and σ_{I_i} are the mean and standard deviation, respectively, of time series data "I_i". At the end of the training process, the outputs are denormalized into the original data format for obtaining the desired outputs.

In this way, the remaining neural networks (BP2–BP5) are trained and tested separately for June, July, August, September and seasonal time series data by subsequently increasing the number of nodes in the input and hidden layers. As the number of nodes in the input and hidden layers are increased successively in our proposed neural networks, we also need to increase the number of observed rainfall values (X_i, X_j, X_k, X_l and X_m) in the input arrays 7.4.2–7.4.6 successively.

7.5 Ensemble of Outputs

Due to handling of large number of input variables, hidden neurons and additional parameters, the ANN outputs tend to become unstable. To resolve the problem of instability, an ensemble neural network approach is adopted (Perrone and Cooper 1993; Cannon and McKendry 1999). In this approach, outputs obtained from various neural networks are combined together to improve the accuracy as well as the stability of the model. In this work, the final output is computed by taking the average of combined outputs from the individual neural networks. Mathematically, output of the ensemble approach can be defined as (Maqsood et al. 2004):

$$O_{final} = \frac{1}{n}\sum_{i=1}^{n} O_i(x), \tag{7.5.1}$$

where $O_i(x)$ is the function computed by the ith neural network, and n is the total number of neural networks trained.

The main aim of employing this approach is for its ease of understanding and implemention (Bishop 1995; Liu et al. 2000), and also its error rate lies within the acceptable range.[4]

[4]References are:(Bishop 1995; Shimshoni and Intrator 1998; Sharkey 1999).

7.6 Simulation Results and Discussions

7.6.1 Empirical Analysis

The main objective of this research is to present a neural network model for predicting ISMR values from existing time series data, so that it can represent the dynamic nature of rainfall. For this purpose, the five neural networks are trained and tested separately using the monthly and seasonal time series data. During their learning process, different experiments were carried out to set the values of additional parameters (see Sect. 7.4.1) to obtain the optimal results, and we have chosen the ones that exhibit the best behavior in terms of accuracy. In this work, the *initial weight*, *learning rate* and *momentum* are taken as 0.3, 0.5 and 0.6, respectively. The parameter *minimum weight delta* is also adjusted to speed up the learning process and it is set to 0.0001.

We compare our results with that of Sahai et al. (2000) model. Both competing model and the proposed model use the BPNN algorithm, which is more sensitive to the number of neurons. A few numbers of neurons can lead to underfitting, while large numbers can contribute to overfitting (Demuth et al 2008a). In the proposed neural networks, the number of neurons in the input layer vary from 5 to 9 (see Table 7.3), whereas the Sahai et al. (2000) model uses 25 number of neurons in the input layer. Another complexity in the Sahai et al. (2000) neural network architecture is that it consists of four layers (*one input layer, two hidden layers and one output layer*), whereas the proposed neural networks consist of three layers (*one input layer, one hidden layer and one output layer*). Therefore, the proposed neural networks are simple and less complex in comparison to the Sahai et al. (2000) neural network, which can reduce the possibility of overfitting to some extent. Another effective measure is taken in this work to avoid the overfitting problem by stopping the learning process of the neural networks as soon as the error rates reach to an acceptable limit. All these error rates were observed over 10, 000 number of epochs. The predicted results from all the five neural networks are then ensembled together (see formula (7.5.1)).

The performance of the proposed model is evaluated with the help of $\bar{A}$ and SD of the observed and predicted values, R between observed and predicted values, $RMSE$ and PP. These parameters are defined in Chap. 2 (see Sect. 2.6). The statistics of the results obtained for the training data are presented in Table 7.4. The last columns of Table 7.4 depict the results reported by the Sahai et al. (2000) model. Now, a comparison of our statistics of rainfall predictions is made with the results of the Sahai et al. (2000) model. The $\bar{A}$ and SD of the observed and predicted values for the proposed model are very close to that of the actual values in comparison to the Sahai et al. (2000) model. The R values between the actual and predicted rainfall for the four months and season also indicate the efficiency of the proposed model. The forecasting results in terms of $RMSE$ also portray very small error rate in comparison to the Sahai et al. (2000) model. The PP values in Table 7.4 also signify the effectiveness of the proposed model.

Table 7.4 Comparison of experimental results of the proposed model and Sahai et al. model (Sahai et al. 2000) for the training and testing data

	Proposed Model (1876–1960)						Model (1876–1960) (Sahai et al. 2000)				
	Statistics	Seasonal	June	July	August	September	Seasonal	June	July	August	September
	$\bar{A}$ Observed (mm)	855.3	164.5	276.4	242.7	171.7	855.3	164.5	276.4	242.7	171.7
	$\bar{A}$ Predicted (mm)	852.4	161.01	272.4	241.54	168.6	858.8	158.3	270.9	240.1	164.7
	SD Observed (mm)	80.8	37.2	38.3	40.8	39.1	80.8	37.2	38.3	40.8	39.1
Training Data	SD Predicted (mm)	72.4	30.3	31.4	35.2	32.5	64.8	23.3	28.2	33.1	28.4
	R	0.92	0.89	0.88	0.92	0.90	0.91	0.85	0.84	0.91	0.82
	$RMSE$ (mm)	28.12	17.3	16.01	14.1	18.08	34.94	22.34	21.94	17.84	23.95
	PP	0.65	0.53	0.58	0.66	0.54	0.18	0.36	0.33	0.19	0.37
	Proposed Model (1961–2010)						Model (1961–1994) (Sahai et al. 2000)				
	Statistics	Seasonal	June	July	August	September	Seasonal	June	July	August	September
	$\bar{A}$ Observed (mm)	833.03	160.76	263.57	242.91	165.79	840.0	157.6	267.1	249.4	166.2
	$\bar{A}$ Predicted (mm)	825.95	157.06	260.01	239.56	163.02	829.5	159.0	269.6	246.5	165.8
	SD Observed (mm)	86.70	34.37	37.21	32.80	34.06	90.4	32.9	33.6	32.0	36.0
Testing Data	SD Predicted (mm)	78.56	30.11	32.01	27.9	28.01	69.2	18.8	22.6	24.0	17.2
	R	0.85	0.87	0.89	0.91	0.88	0.81	0.67	0.63	0.83	0.74
	$RMSE$ (mm)	26.02	18.50	20.09	15.12	20.00	54.24	24.76	26.14	18.33	26.02
	PP	0.70	0.46	0.46	0.54	0.41	0.36	0.56	0.60	0.33	0.52

The prediction results obtained for the testing data are also depicted in Table 7.4 in terms of various statistical parameters. These results are then compared with the results reported by the Sahai et al. (2000) model (last columns of Table 7.4). Though, the range of testing data in this study is larger than the data used by the Sahai et al. (2000) model, but the results clearly exhibit superiority of our model. Table 7.4 states that Sahai et al. (2000) model performs well in training data, but the performance goes down for testing data. This can be conceived from the large fluctuation in the SD of observed and predicted values of Sahai et al. (2000). However, in case of the proposed model, the SD of observed and predicted values are very close to each other. In terms of R, $RMSE$ and PP, the proposed model also exhibits better performance than the Sahai et al. (2000) model.

The comparison graphs are also plotted for the training and testing data as shown in Figs. 7.6, 7.7, 7.8, 7.9 and 7.10, respectively. These figures clearly demonstrate that our model predicts rainfall values, which are very close to observed rainfall values. From the above discussions, it is quite apparent that the proposed model is much better than the Sahai et al. (2000) model in terms of accuracy and performance.

7.6.2 Seasonal Rainfall Prediction: Interpretation in Terms of Hydrology

Due to the dynamic nature of seasonal rainfall (June to September), it's advance prediction with exact amount is a very challenging task. These fluctuations in the quantity of seasonal rainfall over different parts of the country shows significant effect on agriculture and economy. Various definitions have been used to define these fluctuations of rainfall in terms of drought and flood, which are complex hydrological events, and are characterized by a few correlated random variables (Mishra and Singh 2011). Meteorological drought is usually measured by how far from normal the precipitation has been over some period of time (IFAS 1998). In 1971, IMD considered drought to have occurred in a year over a region or subdivision when the seasonal rainfall was less than 75 % of the normal (Pathasarathy et al. 1994). Rajeevan et al. (2004) categorized the seasonal rainfall amount into five categories. Based on these categories, a seasonal rainfall amount which is characterized by less than 90 % of LPA of the seasonal rainfall data is considered as drought.

From the above discussions, it is obvious that it affects surface water as well as ground water, which can lead to reduced water supply, deteriorated water quality, crop failure and disturbed riparian habitats (Travis et al. 1991). It also causes degradation of soil and which leads to desertification (Nicholson et al. 1990; Pickup 1998; Bacanli et al. 2009). Therefore, drought monitoring, drought prediction and analysis of spatial extent of drought risk are the theme of many studies globally in order to assist agricultural or environmental management (Dracup et al. 1980; Wilhite 1996).

Drought prediction is especially important in case of India as it's agriculture heavily relies on natural rainfall which is mainly concentrated in a short span of

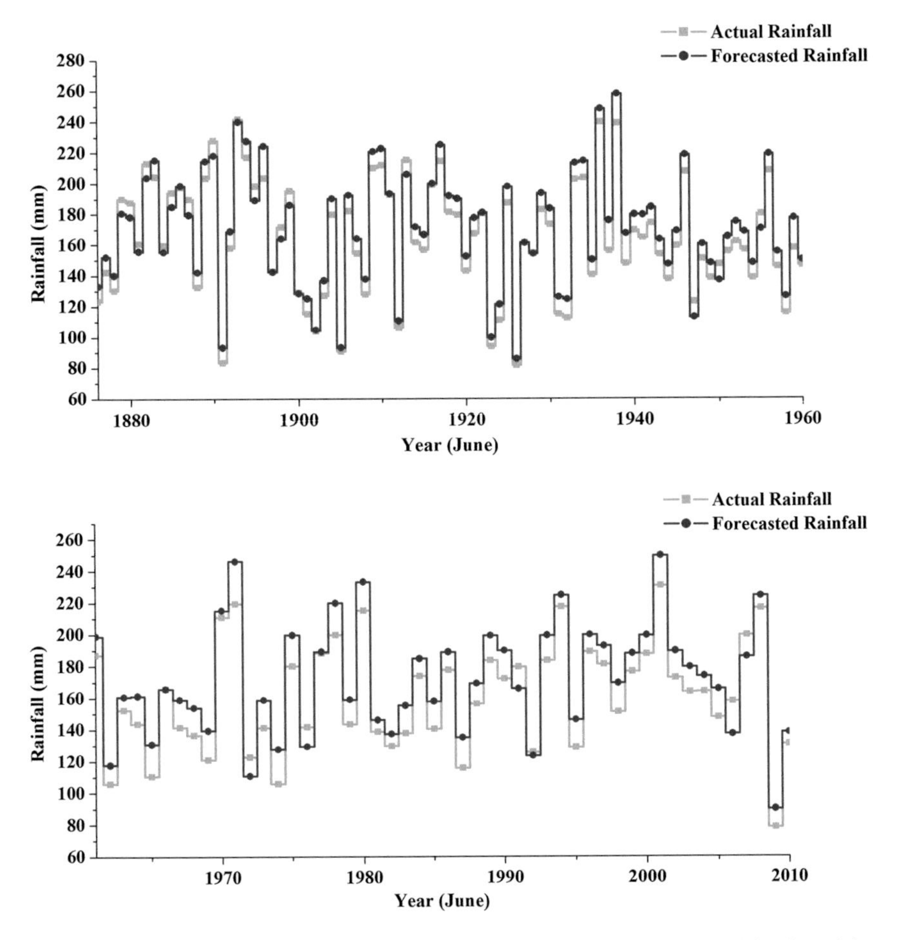

Fig. 7.6 Comparison of observed and predicted June rainfall values (*top* to *bottom*) for the training (1876–1960) and testing (1961–2010) data

time during monsoon season. Any anomaly in this could result in a catastrophe. Many studies have been focused on droughts for small regions or for whole of India. Some researchers have concentrated on the probabilistic study of drought (Chowdhury and Abhyankar 1984; Krishna et al. 1984), whereas some have focussed on the climatological aspect of flood (Chowdhury and Abhyankar 1984). A statistical approach has been used by Chowdhury et al. (1989) to study drought incidences over India. Shewale and Ray (2001) worked on probabilities of occurrence of drought in various sub-divisions of India. Gore and Ray (2002a, b) studied droughts over Maharashtra (2002) and Gujarat (2002, 2004) over smaller spatial scale.

Most recently, new techniques such as ANN, FL and hybridization of ANN and FL have been applied as an efficient alternative tool for modeling complex hydrologic systems and widely used for forecasting. Many researchers (Jeong and Kim 2005; Kumar et al. 2005; Rajurkarand et al. 2004) utilized ANN for modeling rainfall-runoff

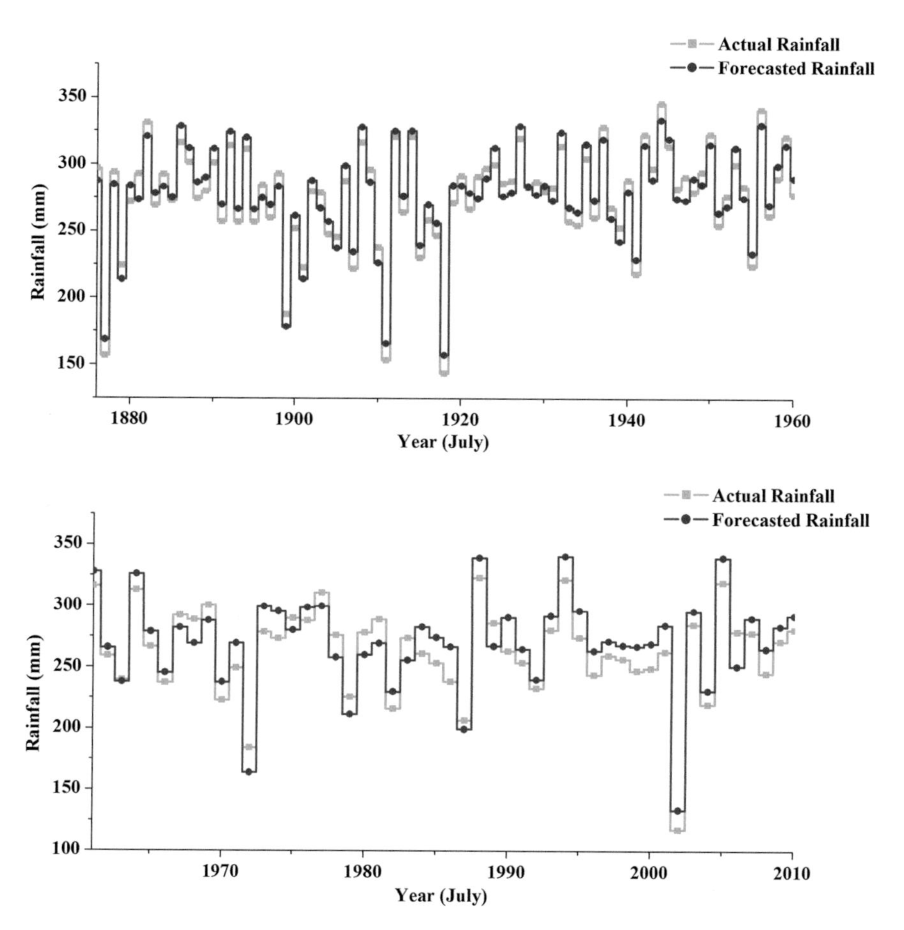

Fig. 7.7 Comparison of observed and predicted July rainfall values (*top* to *bottom*) for the training (1876–1960) and testing (1961–2010) data

process; Jain and Kumar (2007) for hydrologic time series modeling. Medium and long-term prediction of both the likelihood of drought events and their severity has been made using ANN technique (Morid et al. 2007). Mishra et al. (2008) applied the feed-forward recursive neural network and ARIMA models for drought forecasting using SPI series as drought index. The results have demonstrated that neural network method can be successfully applied for drought forecasting.

To illustrate the applicability of the proposed model, the seasonal rainfall amounts for 5 years (2011–2015) in advance have been predicted, which is presented in Table 7.5. According to IMD, the rainfall values are categorized into five categories (IMD 2012). These are: deficient (less than 90 % of LPA), below normal (90–96 % of LPA), normal (96–104 % of LPA), above normal (104–110 % of LPA) and excess (above 110 % of LPA). Based on these categories, the predicted seasonal rainfall amounts are categorized, which are shown in the last column of Table 7.5.

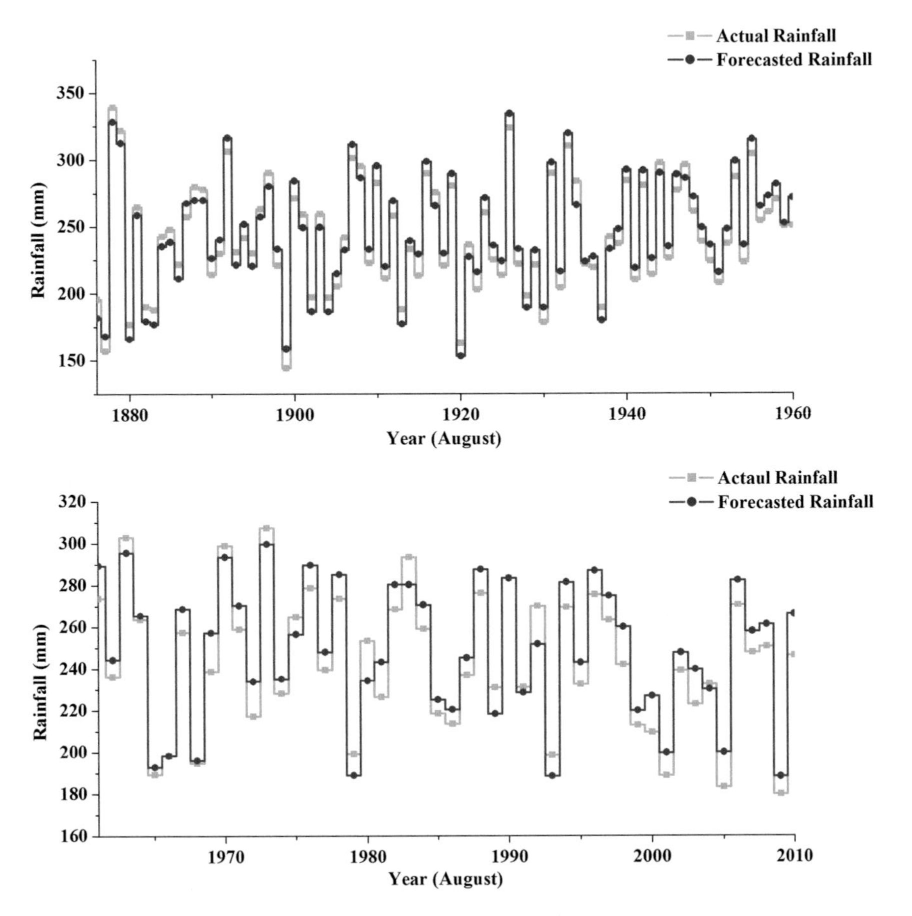

Fig. 7.8 Comparison of observed and predicted August rainfall values (*top* to *bottom*) for the training (1876–1960) and testing (1961–2010) data

According to the reports published by IMD in 2011 and 2012, all India monsoon season rainfall over the country as a whole during 2011 and 2012 were 102 % and 92 % of LPA, which are normal and below normal respectively (Tyagi and Pai 2012; IMD 2012). The proposed model also successfully predicted the seasonal rainfall amounts for 2011 and 2012 as 841.36 mm (103 % of LPA based on 1961–2010 seasonal data) and 831.22 mm (103 % of LPA based on 1961–2010 seasonal data) respectively, that is very close to actual.

The categories of the rainfall as defined above depend on the range of the data selected for computing LPA. Therefore, the predicted category for the seasonal rainfall of 2012 differs from the actual category since we have determined LPA based on the seasonal data of 1961–2010 while LPA value of IMD is based on data of 1951–2000. The proposed prediction of the seasonal rainfall for the remaining years (2013–2015), may however be checked by the successive reports as published by IMD in the current years.

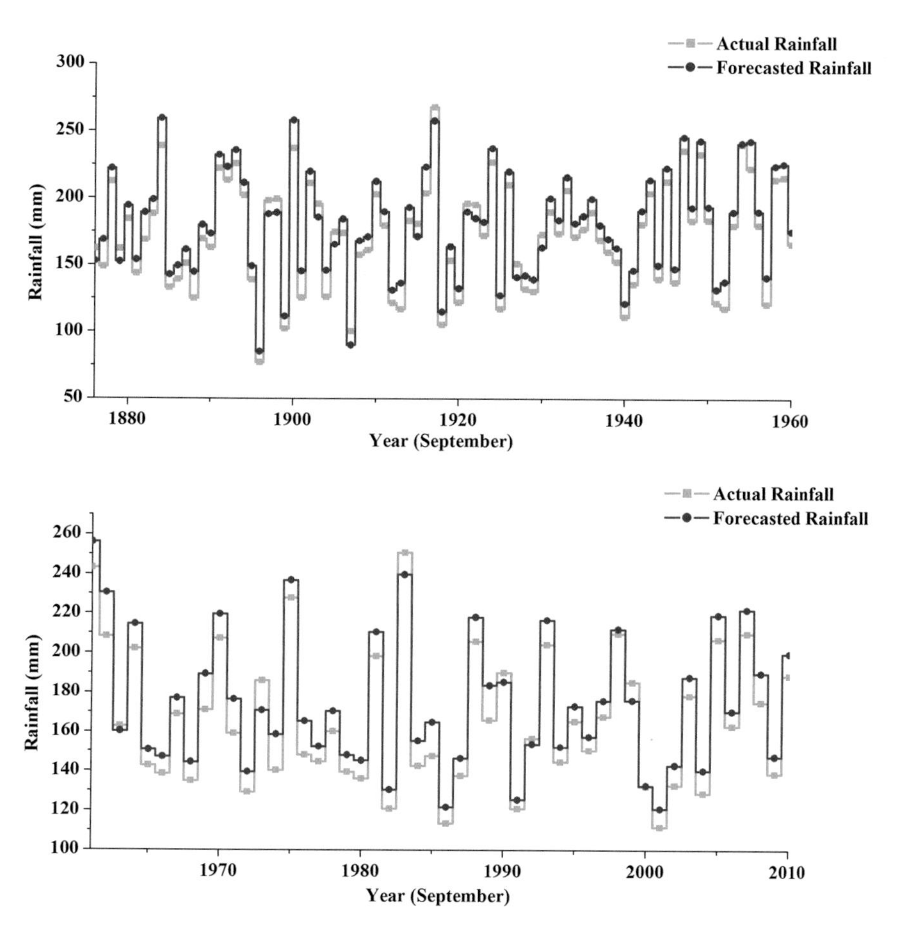

Fig. 7.9 Comparison of observed and predicted September rainfall values (*top* to *bottom*) for the training (1876–1960) and testing (1961–2010) data

7.7 Discussion

The main motivation of the present work is not only to develop a model for the prediction of ISMR on monthly and seasonal time scales, but also to demonstrate it's applicability for advance prediction of the seasonal rainfall amounts. But, the non-stationary nature of ISMR makes it's deterministic prediction more complex. However, various recent studies suggested that the predictability of ISMR in monthly as well as seasonal time scale is possible using time series analysis.[5] Various researchers suggested that dynamic global models have poor skills in predicting ISMR.[6] All these

[5]References are: (Iyengar and Raghukanth 2003; Kishtawal et al. 2003; Rajeevan et al. 2004; Prasad et al. 2010; Kumar et al. 2012; Sinha et al 2012; Singh et al. 2012).

[6]References are: (Kang et al. 2004; Wang et al 2005; Barnston et al. 2010; Sinha et al 2012).

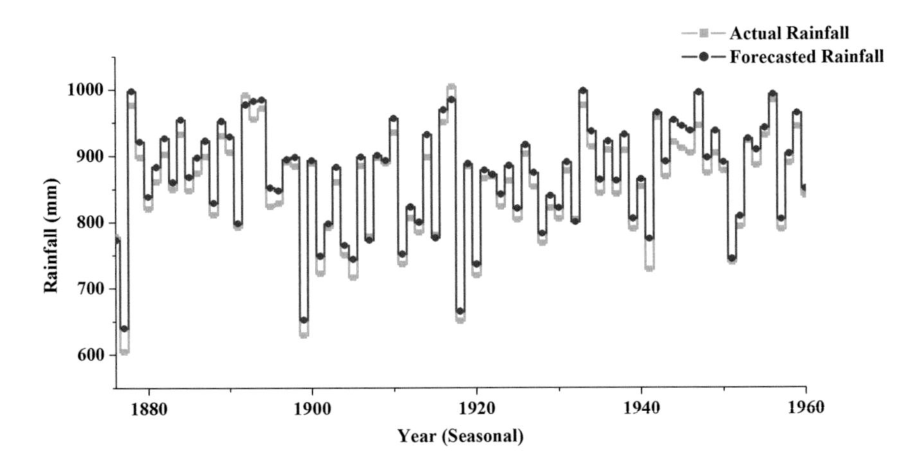

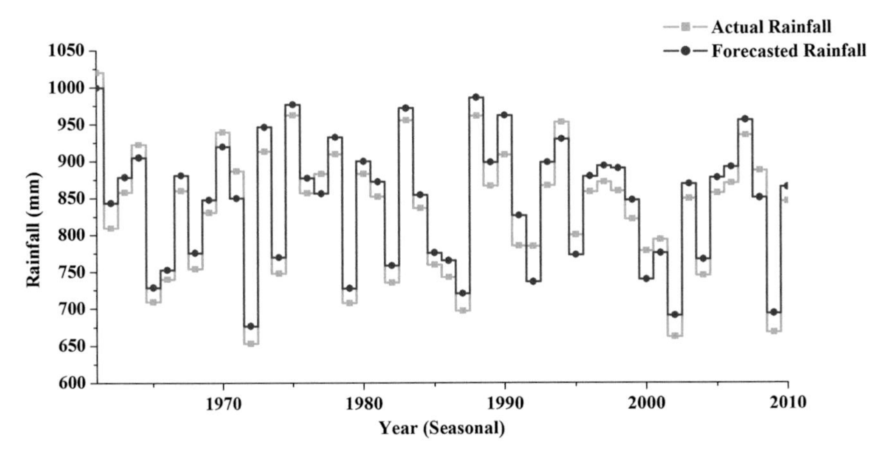

Fig. 7.10 Comparison of observed and predicted Seasonal rainfall values (*left* to *right*) for the training (1876–1960) and testing (1961–2010) data

Table 7.5 Five years advance prediction of the seasonal rainfall amounts (in mm) for the period 2011–2015

Year	Actual rainfall	Category	Predicted rainfall	Category
2011	901.2	Normal	841.36	Normal
2012	819.8	Below normal	831.22	Normal
2013	–	–	814.48	Normal
2014	–	–	800.94	Normal
2015	–	–	826.94	Normal

models have downside that they depend on the interrelationship of global variables or predictors, which changes with time.[7]

The recent studies revealed that the prediction problem of ISMR can be resolved using time series data analysis and ANNs (Chattopadhyay and Chattopadhyay 2001; Chakraverty and Gupta 2008). Therefore, for prediction of ISMR, we have adopted only time series data, and trained and tested the whole data (1871–2010) in the proposed neural network model, and predicted results were obtained. For training and testing purpose, the standard BPNN is employed, which has several advantages over other neural networks such as RBF (Kumar 2011) and PNN (Sivanandam et al 2012). RBF neural network requires thousand of epochs for its learning process, so it is slower than BPNN. RBF neural network also needs more number of neurons for its training (Beale et al. 2010), because the number of neurons in the first layer is determined by the number of input/target pairs in the training data set (Demuth et al 2008b). Another advantage of BPNN over PNN is that we can manipulate the stopping criteria according to the training condition (Chong and Sundaraj 2009).

Results that are obtained are validated using various statistical parameters, which indicate that the presented model is computationally robust and capable of learning very fast. The architecture of the developed networks are comparatively simpler and they are able to predict the non-linear behavior of ISMR much more accurately compared to the Sahai et al. (2000) model. Based on the proposed model, the seasonal rainfall amounts for the next five years have also been predicted. This analysis could be advantageous for advance prediction of events like drought and flood. However, ability of our model to predict rainfall amount in advance can be evaluated by the future years' observations, which will draw more interesting findings and conclusions.

The proposed model is found to be very time efficient for simulation, training, testing, and analyzing the data, which is important from the perspective of prediction studies which involves the prediction of dynamic variables of the environment.

References

Aksoy H, Dahamsheh A (2008) Artificial neural network models for forecasting monthly precipitation in Jordan. Stoch Environ Res Risk Assess 23:917–931

Annamalai H (1995) Intrinsic problems in the seasonal prediction of the Indian summer monsoon rainfall. Meteorol Atmos Phys 55:61–76

Bacanli UG, Firat M, Dikbas F (2009) Adaptive neuro-fuzzy inference system for drought forecasting. Stoch Env Res Risk Assess 23:1143–1154

Barnston AG et al (2010) Verification of the first 11 years of IRI's seasonal climate forecasts. J Appl Meteorol Climatol 49:493–520

Basu S, Andharia HI (1992) The chaotic time-series of Indian monsoon rainfall and its prediction. Proc Ind Acad Sci (Earth Planet Sci) 101:27–34

Beale MH, Hagan MT, Demuth HB (2010) Neural network Toolbox 7. The MathWorks Inc, Natick

Bishop CM (1994) Neural networks and their applications. Rev Sci Instrum 65(6):1803–1832

Bishop CM (1995) Neural networks for pattern recognition. Oxford University Press, Oxford, UK

[7]References are: (Mooley and Munot 1993; Kumar et al 1999; Singh et al. 2012).

Blanford HF (1884) On the connection of the Himalayan snow with dry winds and seasons of droughts in India. Proc R Soc Lond 3–22

Cannon AJ, McKendry IG (1999) Forecasting all-India summer monsoon rainfall using regional circulation principal components: a comparison between neural network and multiple regression models. Int J Climatol 19(14):1561–1578

Carpenter WC, Barthelemy JF (1994) Common misconceptions about neural networks as approximators. ASCE J Comput Civil Eng 8:345–358

Chakraverty S, Gupta P (2007) Comparison of neural network configurations in the long-range forecast of southwest monsoon rainfall over India. Neural Comput Appl 17:187–192

Chakraverty S, Gupta P (2008) Comparison of neural network configurations in the long-range forecast of southwest monsoon rainfall over India. Neural Comput Appl 17:187–192

Chattopadhyay S, Chattopadhyay G (2001) Identification of the best hidden layer size for threelayered neural net in predicting monsoon rainfall in India. J Hydroinform 10(2):181–188

Chong Y, Sundaraj K (2009) A study of back-propagation and radial basis neural network on EMG signal classification. In: 6th International symposium on mechatronics and its applications, Sharjah, UAE, pp 1–6

Chowdhury A, Abhyankar VP (1984) On some climatological aspects of droughts in India. Mausam 35(3):375–378

Chowdhury A, Dandekar MM, Raut PS (1989) Variability of drought incidence over India. A statistical approach. Mausam 40(2):207–214

Demuth H, Beale M, Hagan M (2008a) Neural network Toolbox 6 User's Guide, chap Backpropagation, pp 156–227

Demuth H, Beale M, Hagan M (2008b) Neural network Toolbox 6 User's Guide, chap Radial basis networks, pp 293–307

Dracup JA, Lee KS, Paulson EG (1980) On the statistical characteristics of drought events. Water Resour Res 16:289–296

Gadgil S, Srinivasan J, Nanjundiah RS, Kumar KK, Munot AA, Kumar KR (2002) On forecasting the Indian summer monsoon: the intriguing season of 2002. Curr Sci 83(4):394–403

Gadgil S, Rajeevan M, Nanjundiah R (2005) Monsoon prediction-Why yet another failure? Curr Sci 88(9):1389–1400

Gore PG, Ray KS (2002a) Droughts and aridity over districts of Gujarat. J Agrometeorol 4(1):75–85

Gore PG, Ray KS (2002b) Variability of drought incidence over districts of Maharashtra. Mausam 53(4):533–538

Goswami P, Kumar P (1997) Experimental annual forecast of all-India mean summer monsoon rainfall for 1997 using a neural network model. Curr Sci 72:781–782

Goswami P, Srividya, (1996) A novel neural network design for long range prediction of rainfall pattern. Curr Sci 70:447–457

Guhathakurta P (2006) Long-range monsoon rainfall prediction of 2005 for the districts and subdivision kerala with artificial neural network. Curr Sci 90:773–779

Guhathakurta P, Rajeevan M, Thapliyal V (1999) Long range forecasting Indian summer monsoon rainfall by a hybrid principal component neural network model. Meteorol Atmos Phys 71:255–266

Han J, Kamber M (2001) Data mining: concepts and techniques, 1st edn. Morgan Kaufmann Publishers, USA

Hastenrath S (1988) Prediction of Indian monsoon rainfall: further exploration. J Clim 1:298–304

Hastenrath S, Greischar L (1993) Changing predictability of Indian monsoon rainfall anomalies? Earth Planet Sci 102:35–47

IFAS (1998) Extreme heat and drought. The disaster handbook. University of Florida, Gainesville

IITM (2012) Homogeneous Indian monthly rainfall data sets. http://www.tropmet.res.in/

IMD (2012) 2012 Southwest monsoon end of-season report. Technical report, National Climate Centre, India Meteorological Department, Pune–5, India

Iyengar RN, Raghukanth STG (2003) Empirical modelling and forecasting of Indian monsoon rainfall. Curr Sci 85(8):1189–1201

Jain A, Kumar AM (2007) Hybrid neural network models for hydrologic time series forecasting. Appl Soft Comput 7:585–592
Jeong D, Kim Y (2005) Rainfall-runoff models using artificial neural networks for ensemble streamflow prediction. Hydrol Process 19(19):3819–3835
Johnson RA, Wichern DW (1992) Applied multivariate statistical analysis, 5th edn. Prentice-Hall, New Jersey
Kang IS, Lee JY, Park CK (2004) Potential predictability of summer mean precipitation in a dynamical seasonal prediction system with systematic error correction. J Clim 17:834–844
Khan MR, Ondrusek C (2000) Short-term electric demand prognosis using artificial neural networks. J Electr Eng 51:296–300
Kishtawal CM, Basu S, Patadia F, Thapliyal PK (2003) Forecasting summer rainfall over india using genetic algorithm. Geophys Res Lett 30(23):2203
Krishna KK, Soman MK, Kumar KR (1995) Seasonal forecasting of Indian summer monsoon rainfall: a review. Weather 50:449–467
Krishna YSR, Sastri ASRSA, Rao GGSN, Rao BVR (1984) On prediction of droughts in the Indian Arid region. Mausam 35(3):349–354
Krishnamurti TN, Bedi HS, Subramaniam M (1990) The summer monsoon of 1988. Meteorol Atmos Phys 42:19–37
Kumar A et al (2012) Multi-model ensemble (MME) prediction of rainfall using neural networks during monsoon season in India. Meteorol Appl 19(2):161–169
Kumar ARS et al (2005) Rainfall-runoff modelling using artificial neural networks: comparison of network types. Hydrol Process 19(6):1277–1291
Kumar KK, Rajagopalan B, Cane MA (1999) On the weakening relationship between the Indian monsoon and ENSO. Science 284(5423):2156–2159
Kumar S (2011) Neural networks. Tata McGraw-Hill, New Delhi
Lippmann PR (1989) Pattern classification using neural networks. IEEE Commun Mag 11:47–54
Liu Y, Yao X, Higuchi T (2000) Evolutionary ensembles with negative correlation learning. IEEE Trans Evol Comput 4(4):380–387
Maqsood I, Khan R, Abraham A (2004) An ensemble of neural networks for weather forecasting. Neural Comput Appl 13(2):112–122
Masters T (1993) Practical neural network recipes in C++. Academic Press, New York
MATLAB (2006) Version 7.2 (R2006). http://www.mathworks.com/
Mishra AK, Singh VP (2011) Drought modeling : a review. J Hydrol 403(1–2):157–175
Mishra AK, Singh VP, Desai V (2008) Drought characterization: a probabilistic approach. Stoch Environ Res Risk Assess 23(1):41–55
Mooley DA, Munot AA (1993) Variation in the relationship of the Indian summer monsoon with global factors. Earth Planet Sci 102:89–104
Mooley DA, Parthasarathy B (1984) Fluctuations in all-India summer monsoon rainfall during 1871–1978. Clim Change 6:287–301
Morid S, Smakhtin V, Bagherzadeh K (2007) Drought forecasting using artificial neural networks and time series of drought indices. Int J Climatol 27:2103–2111
Navone HD, Ceccatto HA (1994) Predicting Indian monsoon rainfall: a neural network approach. Clim Dyn 10:305–312
Nicholson SE, Davenport ML, Malo AR (1990) A comparison of the vegetation response to rainfall in the Sahel and East Africa, using normalized difference vegetation index from NOAA AVHRR. Clim Change 17:209–241
Pandya AS, Macy RB (1996) Pattern recognition with neural networks in C++. CRC Press Inc, New York
Parthasarathy B, Rupakumar K, Kothawale DR (1992) Indian summer monsoon ramfall indices: 1871–1990. Meteorol Mag 12:174–186
Pathasarathy B, Munot AA, Kothawale DR (1994) All India monthly and seasonal rainfall series: 1871–1993. Theor Appl Climatol 49:217–224

Perrone MP, Cooper LN (1993) When networks disagree: Ensemble methods for hybrid neural networks. Chapman and Hall, London, pp 126–142
Pickup G (1998) Desertification and climate change-the Australian perspective. Clim Res 11:51–63
Prasad K, Dash SK, Mohanty UC (2010) A logistic regression approach for monthly rainfall forecasts in meteorological subdivisions of India based on DEMETER retrospective forecasts. Int J Climatol 30(10):1577–1588
Preethi B, Revadekar JV, Kripalani RH (2011) Anomalous behaviour of the indian summer monsoon 2009. J Earth Syst Sci 120(5):783–794
Rajeevan M, Pai DS, Dikshit SK, Kelkar RR (2004) IMD's new operational models for long-range forecast of southwest monsoon rainfall over India and their verification for 2003. Curr Sci 86:422–431
Rajurkarand M, Kothyari U, Chaube U (2004) Modeling of the daily rainfall-runoff relationship with artificial neural network. J Hydrol 285(1–4):96–113
Sahai AK, Soman MK, Satyan V (2000) All India summer monsoon rainfall prediction using an artificial neural network. Clim Dyn 16:291–302
Satyan V (1988) Is there an attractor for the Indian summer monsoon? Proc Ind Acad Sci (Earth Planet Sci) 97:49–52
Sharkey AJ (ed) (1999) Combining artificial neural nets: ensemble and modular multi-net systems. Springer, Berlin
Shewale MP, Ray KCS (2001) Probability of occurrence of drought in various sub divisions of India. Mausam 52(3):541–546
Shimshoni Y, Intrator N (1998) Classification of seismic signals by integrating ensembles of neural networks. IEEE Trans Signal Process 46(5):1194–1201
Shukla J, Mooley DA (1987) Empirical prediction of the summer monsoon rainfall over India. Mon Weather Rev 115:695–703
Singh A, Kulkarni M, Mohanty UC, Kar S, Robertson AW, Mishra G (2012) Prediction of Indian summer monsoon rainfall (ISMR) using canonical correlation analysis of global circulation model products. Meteorol Appl 19(2):179–188
Singh P, Borah B (2013) Indian summer monsoon rainfall prediction using artificial neural network. Stoch Env Res Risk Assess 27(7):1585–1599
Sinha P et al (2012) Seasonal prediction of the Indian summer monsoon rainfall using canonical correlation analysis of the NCMRWF global model products. Int J Climatol
Sivanandam SN, Deepa SN (2007) Principles of soft computing. Wiley India (P) Ltd., New Delhi
Sivanandam SN, Sumathi S, Deepa SN (2012) Introduction to neural networks using Matlab 6.0. Tata McGraw-Hill, New Delhi
Swaminathan MS (1998) Padma Bhusan Prof. P. Koteswaram first memorial lecture-23rd March 1998. In: Climate and sustainable food security, vol 28, Vayu Mandal, pp 3–10
Travis WR, Karl T, Changnon SA (1991) Drought and natural resources management in the United States : impacts and implications of the 1987–89 drought. Westview Press, Boulder CO
Tyagi A, Pai DS (2012) Monsoon 2011. Technical report, National Climate Centre, India Meteorological Department, Pune–5, India
Walker GT (1933) Seasonal weather and its prediction. Brit Assoc Adv Sci 103:25–44
Wang B, Ding Q, Fu X, Kang IS, Jin K, Shukla J, Doblas-Reyes F (2005) Fundamental challenge in simulation and prediction of summer monsoon rainfall. Geophys Res Lett 32(L15711)
Wilhite DA (1996) A methodology for drought preparedness. Nat Hazards 13:229–252
Wu X et al (2008) Top 10 algorithms in data mining. Knowl Inf Syst 14:1–37

Chapter 8
Conclusions

Everything should be made as simple as possible, but not simpler.

By Albert Einstein (1879–1955)

Abstract The final chapter of the book concludes (a) the contributions in the domain (refer to Sect. 8.1), and (b) those future research works that are associated with the domain, which require further investigations by the scientific community (refer to Sect. 8.2).

8.1 Contributions

The main motivation for the research work is the growing need of time series forecasting in nearly all fields of natural and social sciences and engineering, especially weather and financial forecasting. However, the main problem in the time series forecasting is to choose the best methodology for fulfilling the desired goals and objectives under reasonable time. To deal with these issues, an extensive literature reviews were carried out and it was concluded that out of the various methodologies, SC is the most appropriate and efficient technique for resolving these. SC is the amalgamated domain of different methodologies such as fuzzy sets, ANN, EC, rough sets and probabilistic computing. Therefore, among these techniques, to find which technique is most suitable for our problem, consumed a lot of time. As most of data under prediction are uncertain in nature, it may represent the past behavior of the system, but it unable to predict the future behavior. So to handle these uncertainties in the data, fuzzy sets theory is the most appropriate one. Song and Chissom (1993) used this theory in the forecasting of time series and popularly named as "FTS forecasting model". Motivated from this, we have extended their idea in the present book on resolving various domain specific problems based on time series forecasting. Here, we have reported four significant contributions in time series forecasting models using SC techniques (especially FTS) and their hybridization. Apart from

P. Singh, *Applications of Soft Computing in Time Series Forecasting*,
Studies in Fuzziness and Soft Computing 330, DOI 10.1007/978-3-319-26293-2_8

these four major contributions, this book also reported one research work based on application of ANN technique in time series forecasting.

The main contributions of the book are summarized as follows. Five different time series forecasting models using SC techniques are introduced:

1. A basic single-factor FTS forecasting model.
2. A single-factor high-order fuzzy-neuro hybridized time series forecasting model.
3. A two-factors high-order neuro-fuzzy hybridized forecasting model.
4. A FTS-PSO hybridized model for M-factors time series forecasting.
5. An ANN based Indian Summer Monsoon rainfall prediction technique.

The first model proposed is an improvement over the original works presented by Song and Chissom (1993), and later modification by Chen (1996). We have contributed a new data discretization approach entitled as "MBD". In this model repeated FLRs are given due weightage, which was the limitation of previous existing FTS models. To define weight for each FLR, a new approach entitled as "IBWT" is incorporated. It is also recommended to employ the "IBDT" for defuzzification operation.

Many researchers suggested that high-order FLRs improve the forecasting accuracy of the models. Therefore, in the development of second model, the high-order FLRs is used to obtain the forecasting results. In this model, for creating the effective lengths of intervals of the historical time series data set, a new "RPD" approach (refer to Sect. 4.4) is introduced. We have also suggested the use of previous state's fuzzified values (in the left hand side of FLRs). To obtain the final forecasting results from these FLRs, an ANN based architecture (refer to Fig. 4.1) is incorporated in the proposed model. The effectiveness of this model is established over real life data sets.

In real-time, one observation (variable) always relies on several observations. A new model is developed to deal with the forecasting problems of two-factors time series data. The proposed model is designed by hybridizing ANN with FTS. This model uses the ANN for clustering the time series data set into different groups. We have also introduced some rules for interval weighing (refer to Sect. 5.3.6) to defuzzify the fuzzified time series data sets. This model has been established to be effective in forecasting the time series with optimal number of intervals.

To improve the forecasting accuracy, all the observations can be incorporated in the forecasting model. Therefore, a new Type-2 FTS model which can utilize more observations in forecasting is introduced. The predictability of this Type-2 model is later enhanced by employing PSO technique. The main motive behind the utilization of the PSO with the Type-2 model is to adjust the lengths of intervals in the universe of discourse that are employed in forecasting, without increasing the number of intervals. The daily stock index price data set of SBI is used to evaluate the performance of the proposed model. The proposed model is also validated by forecasting the daily stock index price of Google.

Lastly, an attempt is made to demonstrate the application of ANN in ISMR forecasting. For this purpose, we have used BPNN algorithm, and presented five neural network architectures designated as BP1, BP2, ..., BP5 using three layers of neurons (one input layer, one hidden layer and one output layer). Separate data set are used for training as well as testing each of the neural network architectures, *viz.*,

BP1–BP5. The forecasted results obtained for the training and testing data are then compared with existing model. Results clearly exhibit superiority of our model over the considered existing models. In this study, the seasonal rainfall values over India for next 5 years have also been predicted.

From the above discussion, following general points can be concluded:

- SC techniques fit very well with the time series data sets (especially weather and financial) which are very non-stationary and uncertain in nature.
- Hybridized models (as discussed in Chaps. 4, 5 and 6) are more robust and efficient than non-hybridized model (as discussed in Chap. 3).
- Models with more parameters give better forecasting results as discussed in Chaps. 5 and 6.
- In the book, the proposed models are verified and validated with different weather and financial time series data sets. Empirical analyzes signify that all these models have the robustness to deal the time series data sets very efficiently than various conventional FTS and statistical models (as discussed in the Chaps. 3 and 6).
- The performance of the models are also evaluated with various statistical parameters, which signify the efficiency of the models.

8.2 Future Work

In this book, we introduced various forecasting models based on the SC techniques. However, this study deserve further studies, therefore the final section we would like to suggest a few significant future works closely related to our study.

- In observation of certain event, recorded time series values not only depend on previous values but also on current values. Therefore, representation of FLR in terms of high-order is a worthy idea in FTS modeling approach (Cheng et al. 2006). However, defining FLR in high-order is more complicated and computationally more expensive than first-order (Aladag et al. 2010). There is a need to put more stress on development of new methods that can automatically determine the optimal order of the high-order FLRs to deal with the forecasting problems.
- The multivariate FTS models are based on the prior assumption that one-factor always dependent on other factors. In order to fuzzify all these factors together, it is very much essential to extract the hidden information from the data, and then try to explore the membership values of each datum. To tackle this problem, many researchers use FCM technique.[1] Some researchers[2] introduce unsupervised clustering techniques that determine the membership values efficiently. In spite of

[1]References are: (Aladag et al. 2012; Chen and Chang 2010; Cheng et al. 2008; Li et al. 2008, 2011, 2010).

[2]References are: (Bahrepour et al. 2011; Bang and Lee 2011; Chen and Tanuwijaya 2011; Chen and Wang 2010; Egrioglu et al 2011; Huarng and Yu 2006; Singh and Borah 2013).

all these development, there is the need for future research on developing more robust data clustering algorithm for multivariate FTS model.

- The FTS models should consider the change in trend associated with the time series in terms of upward, downward or unchanged, besides predicting the future values. Development of more robust trend-based models can be tried.
- This study reflects that hybridized models are more robust than conventional FTS models. However, difficulties arise in determining the applications of such techniques in suitable phase. Therefore, there is the need to develop model selection techniques that can effectively make the use of both input variables and knowledge, and fulfill the forecasting objectives.
- Most of the existing FTS models have used Chen's defuzzification method (Chen 1996) to acquire the forecasting results. However, forecasting accuracy of these models are not good enough. In this book, we also introduced new defuzzification techniques in the Chaps. 3–6. In spite of these contributions, there is scope to propose new defuzzification techniques. Entropy based methods can be tried.

References

Aladag CH, Yolcu U, Egrioglu E (2010) A high order fuzzy time series forecasting model based on adaptive expectation and artificial neural networks. Math Comput Simul 81(4):875–882

Aladag CH, Yolcu U, Egrioglu E, Dalar AZ (2012) A new time invariant fuzzy time series forecasting method based on particle swarm optimization. Appl Soft Comput 12(10):3291–3299

Bahrepour M, Akbarzadeh-T MR, Yaghoobi M, Naghibi-S MB (2011) An adaptive ordered fuzzy time series with application to FOREX. Expert Syst Appl 38(1):475–485

Bang YK, Lee CH (2011) Fuzzy time series prediction using hierarchical clustering algorithms. Expert Syst Appl 38(4):4312–4325

Chen SM (1996) Forecasting enrollments based on fuzzy time series. Fuzzy Sets Syst 81:311–319

Chen SM, Chang YC (2010) Multi-variable fuzzy forecasting based on fuzzy clustering and fuzzy rule interpolation techniques. Inf Sci 180(24):4772–4783

Chen SM, Tanuwijaya K (2011) Multivariate fuzzy forecasting based on fuzzy time series and automatic clustering techniques. Expert Syst Appl 38(8):10594–10605

Chen SM, Wang NY (2010) Fuzzy forecasting based on fuzzy-trend logical relationship groups. IEEE Trans Syst Man Cybern Part B: Cybern 40(5):1343–1358

Cheng C, Chang J, Yeh C (2006) Entropy-based and trapezoid fuzzification-based fuzzy time series approaches for forecasting IT project cost. Technol Forecast Soc Chang 73:524–542

Cheng CH, Cheng GW, Wang JW (2008) Multi-attribute fuzzy time series method based on fuzzy clustering. Expert Syst Appl 34:1235–1242

Egrioglu E, Aladag CH, Yolcu U, Uslu V, Erilli N (2011) Fuzzy time series forecasting method based on Gustafson-Kessel fuzzy clustering. Expert Syst Appl 38(8):10355–10357

Huarng K, Yu THK (2006) Ratio-based lengths of intervals to improve fuzzy time series forecasting. IEEE Trans Syst Man Cybern Part B: Cybern 36(2):328–340

Li ST, Cheng YC, Lin SY (2008) A FCM-based deterministic forecasting model for fuzzy time series. Comput Math Appl 56(12):3052–3063

Li ST, Kuo SC, Cheng YC, Chen CC (2011) A vector forecasting model for fuzzy time series. Appl Soft Comput 11(3):3125–3134

Li ST et al (2010) Deterministic vector long-term forecasting for fuzzy time series. Fuzzy Sets Syst 161(13):1852–1870

Singh P, Borah B (2013) High-order fuzzy-neuro expert system for daily temperature forecasting. Knowl-Based Syst 46:12–21

Song Q, Chissom BS (1993) Forecasting enrollments with fuzzy time series—Part I. Fuzzy Sets Syst 54(1):1–9

Appendix A
Author's Publications

The contribution of author's journal publications in this book are mentioned below:

1. P. Singh and B. Borah. An effective neural network and fuzzy time series-based hybridized model to handle forecasting problems of two factors. *Knowledge and Information Systems (Springer)*, **38**(3), 669–690, 2012.
2. P. Singh and B. Borah. Indian summer monsoon rainfall prediction using artificial neural network. *Stochastic Environmental Research and Risk Assessment (Springer)*, **27**(7), 1585–1599, 2013.
3. P. Singh and B. Borah. An efficient time series forecasting model based on fuzzy time series. *Engineering Applications of Artificial Intelligence (Elsevier)*, **26**, 2443–2457, 2013.
4. P. Singh and B. Borah. High-order fuzzy-neuro expert system for daily temperature forecasting. *Knowledge-Based Systems (Elsevier)*, **46**, 12–21, 2013.
5. P. Singh and B. Borah. Forecasting stock index price based on M-factors fuzzy time series and particle swarm optimization. *International Journal of Approximate Reasoning (Elsevier)*, **55**, 812–833, 2014.
6. P. Singh. A brief review of modeling approaches based on fuzzy time series. *International Journal of Machine Learning and Cybernetics (Springer)*, pages 1–24, 2015, DOI:10.1007/s13042-015-0332-y.

The following references are related to publications of the author in the proceeding of national/international conferences:

1. P. Singh and B. Borah. A multi-purpose forecasting model based On fuzzy time-series. Doctoral Colloquium, IDRBT, Hyderabad, 2011.
2. P. Singh and B. Borah. An efficient method for forecasting using fuzzy time series. Trends in Machine Intelligence, Tezpur University, Assam, 2011.
3. P. Singh and B. Borah. Prediction of all India summer monsoon rainfall using artificial neural network. OCHAMP, IITM, Pune, 2012.

P. Singh, *Applications of Soft Computing in Time Series Forecasting*,
Studies in Fuzziness and Soft Computing 330, DOI 10.1007/978-3-319-26293-2

Index

P. Singh, *Applications of Soft Computing in Time Series Forecasting*,
Studies in Fuzziness and Soft Computing 330, DOI 10.1007/978-3-319-26293-2